Humanising Safety

The world of safety for professionals can often be unclear. In an industry that divides the safety world into one of two camps, either traditional or contemporary safety, the lack of knowledge continues to propagate through books, publications, podcasts, social media, and conferences, leaving safety professionals feeling more muddled than informed.

Humanising Safety: A Four-Step Approach provides a practical approach to human-centric safety collating the best elements of traditional and contemporary approaches for safety professionals to practise at work. By following this approach, readers will learn to apply humanistic safety principles to any workplace where safety is valued. This book explores the realm of human-centric safety and its intricacies, unpacking topics such as the contradictions and dilemmas of workplace safety, the psychology of safety, the human condition and its contribution to the safety of work, and how safety leaders can synthesise the collective knowledge, skills, expertise, and lived experiences of the people who make up an organisation. Featuring micro-projects for readers to refer to and work through within their organisations, this book allows the reader to navigate the vast sea of information surrounding the opportunities and pitfalls of traditional and contemporary safety approaches through a lens of human-centric safety.

This is an easy-to-read book that will appeal to professionals at all career levels where safety is critical to their role, including those in construction, utilities, manufacturing, mining, civil, aviation, and maritime sectors.

Tim D'Ath is a senior safety executive and people leader in Melbourne, Australia. He has more than 15 years of experience leading high-performing teams in both corporate and high-risk operational environments. Prior to this, Tim spent ten years as a construction worker, mobile plant operator, offshore oil rig roustabout, deckhand, and trade assistant. His experience in frontline worker roles has directly shaped his approach to working as a safety professional, acknowledging the skills, experience, and perspectives of frontline workers in the development of safety programs. He specialises in the psychology of safety with extensive experience implementing contemporary safety approaches and psychosocial well-being strategies, and developing health and safety governance frameworks across complex matrix structures. He has worked in diverse industries including construction, maritime, aviation, and utilities, applying humanistic safety approaches in highly regulated work environments.

Humanising Safety
A Four-Step Approach

Tim D'Ath

CRC Press is an imprint of the
Taylor & Francis Group, an **informa** business

Designed cover image: Shutterstock

First edition published 2025
by CRC Press
2385 NW Executive Center Drive, Suite 320, Boca Raton FL 33431

and by CRC Press
4 Park Square, Milton Park, Abingdon, Oxon, OX14 4RN

CRC Press is an imprint of Taylor & Francis Group, LLC

ISBN: 978-1-032-67950-1 (hbk)
ISBN: 978-1-032-66618-1 (pbk)
ISBN: 978-1-032-67952-5 (ebk)

DOI: 10.1201/9781032679525

Typeset in Times
by codeMantra

Contents

Part 3 A Four Step Approach to Humanising Safety

Foreword

I first met the author at a safety conference in Sydney, 13 years ago. I had just finished delivering a masterclass on applying psychology to safety and was eager to have a coffee and some quiet time. This plan was thwarted when Tim approached me with some questions about my session. As I put my laptop away (and surreptitiously looked at my watch), I provided short answers to Tim's very astute questions, hoping my badly needed coffee would not be too far away. But the questions kept coming! And my answers become longer and more considered. There was something about this guy that drew me in.

We ended up having a break together (finally some coffee!) and I learned Tim had recently accepted a safety management role on a multi-billion dollar port expansion project. I remember thinking he seemed young for such a position.

Thirteen years later, I regard Tim as one of the very best safety leaders I have worked with. His insatiable curiosity and willingness to adopt an eclectic approach to his work have seen him excel in industries such as construction, utilities, and aviation.

Tim has experienced safety from both sides of the fence: From an "on the tools" perspective to a senior safety leadership role within a high reliability organisation.

Without a doubt, Tim's blue-collar experience has shaped his approach to safety leadership (and indeed, leadership in general).

Like many of us who have worked at the shop floor level, Tim experienced safety being done TO him, by safety officers, who wielded a plethora of seemingly pointless 'tick and flick' obstacles to getting the job done.

When he moved into safety roles himself, he was wise enough to realise he needed to work WITH teams - not just to bring them on the safety journey with him, but to enable them to actually co-design the journey.

Tim's interest (and studies) in psychology has helped him to formulate a humanistic and eclectic approach to creating a 'safety coalition' with work teams. He understands that safety is not about blind adherence to one modality or approach, regardless of how popular it may be.

For Tim, safety is more about first gaining a thorough understanding (with and from the team members) of the work to be done and then collaboratively drawing from the buffet of available theories and models to formulate a plan.

Not all plans work out! However, many do, and Tim has seen some stunning successes in his career (which no doubt he would credit to his teams, rather than his leadership).

I have been nagging Tim to document these successes and to share what he's learned for a while now. I couldn't be happier that my nagging seems to have finally paid off!

In *Humanising Safety: A Four Step Approach,* Tim offers safety professionals highly practical methods to build a safety coalition. His tone is refreshingly conversational and authentic. There are so many safety leadership books available; many (in fact most) have been written by academics, commentators, or consultants. Few, as in this case, have been written BY practitioners FOR practitioners, and that makes all the difference in this book.

I urge you, dear reader, to note the many micro-projects Tim offers throughout this book. They are a gift and will help you to realise Tim's central goal of humanising safety.

Enjoy!

Clive Lloyd
Psychologist and Author of "Next Generation Safety Leadership: From Compliance to Care".
Director and Principal Consultant with GYST Consulting

Preface

Safety. What do you think of when you hear this word? What kind of emotions does it stir? Does it make you feel supported, comforted, present, and in control? Or does it conjure feelings of fear, uncertainty, cynicism, or resentment? Does feeling safe translate into being safe? Or could strong feelings of safety paradoxically compromise our actual safety by allowing complacency to creep in? Does the importance you place on safety vary depending on the context in which it is used? For example, is workplace safety just as important to you as personal safety, emotional safety, financial safety, or cultural safety? Do you value one more than the other? If so, why do you think that might be?

If people were to respond to these questions about safety in the exact same way, then the role of the safety professional would be simple. There would be one safety framework perfectly sufficient to apply to any workplace, regardless of industry or organisational context, and regardless of the people, personalities, beliefs, attitudes, and cultures that make up an organisation.

But this isn't the case. People differ in their responses to questions like these. Their thoughts, feelings, and attitudes around workplace safety are influenced by their lived experiences, what they have been taught about safety, how they perceive or experience safety, their psychological makeup, and the value they place on keeping themselves safe at work. Safety is different for everyone because everyone is different. I don't think it could be said in a simpler way.

So why then, do so many safety professionals continue to engage in these back-and-forth arm wrestles between safety models, giving little regard to the organisational context, industry drivers, and human factors that differ so much between organisations? Safety I, Safety II, traditional safety, the New View, Behavioural-based Safety, Safety Differently, Human and Organisational Performance (HOP), Human Factors. There is gold to be found in *all* these approaches. I don't think it's a matter of picking your preferred model and discarding the others. Workplaces, like people, are all different. And each workplace has different needs when it comes to safety. So, doesn't it make sense to first understand the needs of a workplace and its people, and then develop a safety approach based on those needs, pulling from the generous library of information that exists across all these safety models? It sure makes sense to me.

Unfortunately, workplace safety is surrounded by negative stigma – and it's pretty easy to understand why. Over decades of missteps, the safety profession has managed to take something so inherently human and has proceduralised it to the point that it is no longer about people at all, leaving only a tangled mess of contradictions and a workforce paralysed by indecision.

This goes to the heart of control versus empowerment. Employees are expected to report incidents and near misses and are then blamed through investigation processes. They join organisations that claim safety is a company value, only to find individualistic, nonsensical slogans at every corner, like "safety starts with you". Employees are told they are empowered to intervene in unsafe situations, but when production and safety face-off in a managerial question of priorities, safety almost always takes a back seat. Safety cultures claim to be proactive or generative, but often the workforce trust required for such claims is not there, as safety programs are measured by compliance and driven by fear. Yep, workplace safety is a mess, and I think it's about time we cleaned it up.

I'm not a Safety-I guy. I'm not a Safety-II guy either. I'm just someone who works in the safety profession, trying my best to apply safety methodologies that suit the needs of the organisations I work with and that resonate with their people. Because safety is ultimately about people – the human condition and its contribution to the safety of work. All the factors that influence how people think, feel, and behave in situations that have the capacity to compromise workplace safety, and how cognitive, social, cultural, and behavioural factors influence an individual's perceptions of risk. For me, it's about humanising safety. Helping people relate to safety in ways that are meaningful for them and having the flexibility to adapt my approaches depending on the person, the situation, and the organisation. And this certainly isn't achieved by applying the same, rigid safety approach across every organisation I work with. Everyone is different. Every workplace is different.

My experience in understanding how people relate to safety doesn't just come from working in the safety profession and studying psychology. You see, I came up 'on the tools' as a blue-collar worker where a future in workplace safety could not have been further from my mind. In fact, had you told me in my 20s that I would someday work in safety, I would have laughed in your face!

Across ten years I worked as a labourer, landscaper, steel fixer, concreter, mobile plant operator, and construction supervisor in the civil construction and mining sectors, before moving into the oil and gas industry where I made a living as a rigger, dogman, and roustabout on offshore oil rigs. I then spent a couple of years working as a deckhand on rig tenders and dredging vessels in the far reaches of the north-west shelf of Australia. Most of this work involved various fly in, fly out (FIFO) rosters. I worked 4 and 1 rosters (meaning four weeks on and one week off), 3 and 3 rosters, 4 and 4 rosters, and the gruelling 6 and 4 roster for a Dutch dredging company. That was a tough one – working 14-hour days for 42 days straight before receiving a well-earned four-week break. I was a young fella, chasing the money, and I loved getting my work done. There was just one thing that seemed to keep on getting in the way of my work, and that was the relentless interruption of meaningless paper-safe requirements that all of us knew very well, existed only to cover the backsides of management in the event something went wrong. Inductions, start cards, stop cards, take 5s, personal risk assessments, safety moments, safety toolbox meetings, weekly JSA re-writes – you name it!

I didn't hold safety professionals in very high regard when I worked on the other side of the fence. I didn't trust them. I found their condescending approaches offensive, and I didn't believe in what they were selling or how they were selling it. Yep, I subscribed to the negative stigma surrounding workplace safety early in my career, and it was reinforced every time another safety officer felt the need to give me unsolicited instructions on how to perform the most basic of tasks, like how to bend my knees or how to 'safely' walk down a stairwell.

I became cynical of safety very quickly and it took more than a decade of protesting the bureaucracy of safety impracticalities and the dictators standing behind them before I made the decision to move into the safety field myself. I wanted to introduce the concept of practicality to the safety profession (radical, I know!) and I wanted to show organisations that the unhealthy label workplace safety had in Australia could be dismantled with a few simple changes. Listening to the workforce, respecting their expertise and perspectives, removing obstacles, and allowing them to identify safe ways to conduct their work without getting in their way. Sounds pretty straight forward, right? And that has been my approach for the past 15 years as a safety professional in industries such as civil construction, maritime, oil and gas, aviation, and utilities, working with a range of organisations from small construction firms to high-reliability organisations.

I'm not offering anything particularly new in this book. I'm not proposing yet another model to add to the already swollen safety buffet. I'm just summarising what I have learnt along my journey so far, together with what I have begged, borrowed, and stolen from humanistic safety thinkers before me. In fact, I think it's ridiculous that there is even a market for a book promoting the human element of safety, seeking to disrupt the administrative, paper-safe machine that is the safety industry. But the sad truth is that workplace safety has become lost in the corporate world of bureaucratic policies, procedures, checklists, and behaviourist worldviews, with compliance at the centre and humans tossed aside like an undervalued pickle in a burger. All I'm saying is that it's about time we brought the pickle back.

Throughout this book you will find that I offer some tools and micro-projects for you to consider taking away and working through within your organisations. I have hand-picked these based on the success I have had using them in real workplaces across the construction, mining, oil and gas, maritime, aviation and utilities sectors. Whilst some of my approaches have not been terribly successful over the years, these ones have, and my hope is that by implementing the tools and micro-projects that have worked for me, by the time you've finished reading this book you will have the fundamental elements of a human-centric safety strategy that you can roll-out within your organisation.

So, I hope you get something out of this book, and I hope it encourages you to be curious about workplace safety and human behaviour. Because I truly believe we can positively change, how that word – safety – makes so many people feel when they hear it.

Acknowledgements

Five years ago, Clive Lloyd, a long-time mentor and friend, asked if I would consider writing the foreword for his wonderful book, *Next Generation Safety Leadership – from Compliance to Care*. I was humbled that from the long list of safety leaders Clive had worked with over the years, he asked me to open his book. Honestly, I didn't think I would be much good at writing a foreword but once I put pen to paper (or fingers to keys), I found I actually had quite a lot to say! I think Clive picked up on this too, because fast forward a few years and Clive was fervently encouraging me to write a book of my own, introducing me to his publishers, and providing them with an outrageously generous recommendation of me. So, thank you Clive for pushing me, supporting me, and helping me grow as a leader. Thanks also to CRC Press and the Taylor and Francis Group for publishing my project.

I would also like to thank Josh Bryant, Kym Bancroft, Brad Green, and Zoe Nation for their invaluable insights and generosity in contributing to this book with personal examples of how they have approached and adopted humanistic approaches in their own safety leadership journeys. Josh, Kim, Brad, and Zoe are four safety mavericks I admire deeply, and their contributions to the greater safety profession cannot be understated. They are moving the dial, challenging mindsets and helping safety professionals like myself to do the same. I am incredibly humbled that they agreed to contribute to this book.

Thanks also to safety legend Steve Wintle, who indulged me on numerous occasions, sitting in the car at Port Melbourne, swapping philosophies about safety, work, and everything in between. We had a lot of fun working together and at times it seemed downright ridiculous we were getting paid for it!

Thanks to my wife, Tara, for continually pushing me to do things I am reluctant to do (like returning to study and writing this book) and for being my greatest supporter along the way. Every good man needs a Tara.

Finally, thank you to everyone who has purchased this book. I hope you find something within these pages that you can add to your toolkit and I hope this book encourages you to lead with care and curiosity.

Part 1

What Is Safety?

1

A Definition

It always surprises me that so few safety professionals and leaders I speak with have defined an answer to this seemingly simple question. What is safety in the context of your organisation? I think it's a logical place to start for anyone looking to develop or build on their organisations' safety strategy, yet it is a question rarely explored by safety professionals and their senior leadership teams and therefore not well understood by the people whom workplace safety affects the most – those doing the work. Often, these poor folks are left to elucidate the meaning of safety themselves, making assumptions based on the messaging that trickles down to them on the shop floor and by observing how management responds to things like safety incidents and often well-justified deviations from dogmatic safety procedures. Think about how workers interpret platitudes such as 'Safety First' and 'Zero Harm', or risk-washing statements like 'Safety is *your* responsibility'. Without careful framing and a well-understood organisational definition for safety, these slogans and company positions are unlikely to result in their intended consequences. Instead, they quietly erode trust, build cynicism among the workforce, and even demotivate people when they hear utterings from the safety team or management on the next safety program headed their way.

The safety profession is largely about problem solving, right? Solving problems related to reducing unwanted outcomes, increasing desired outcomes, and finding practical ways for workers to consider health and safety in their everyday work. Most people know that it is difficult to solve a problem for which there is no defined or well-understood problem statement, and a problem statement cannot exist without a half decent explanation of what success looks like.

So, again, what is the definition of safety in the context of your organisation? What does success look like? What does failure look like? Has this been clearly defined? For some, a simple description in the Oxford Dictionary might be enough to get this conversation started. For others, a definition might be informed by their organisations' critical risk controls, worker's perceptions around safety, injury prevention efforts or by the way safety points to, and draws from the organisational strategy. Perhaps it depends on which safety approach best suits your organisation and how that approach frames safety? Or maybe a definition for safety relies on whether it is viewed as an outcome or a process? Whatever the case, it's a great question for any safety professional or leader to ask of their organisation. In fact, I think it's a vital

DOI: 10.1201/9781032679525-2

question and one that needs to be explored early – well before any fruitless attempts to measure safety via KPIs or lead and lag indicators!

Unfortunately, we cannot look to Australian health and safety legislation to provide a definition either. Whilst there are clear legislative obligations related to the provision of safe workplaces, the legislation stops short of providing a definition for safety. It appears that this is a job for you and I and the organisations we work with, which brings us to our first micro-project in building a human-centric safety strategy that you can find at the end of this chapter. But before we get to that, let's compare and contrast the two predominant safety camps for a moment. Just to prime ourselves as we prepare to drift out to meet our first micro-project.

2

Traditional Safety Approaches

I think most people would agree that one of the primary objectives of workplace safety is to prevent employees from getting hurt while at work. This is ultimately why employers recruit safety professionals right? Or at least, that is where it starts. To help create workplace conditions and practices that protect the health and safety of employees, because of an underlying belief that when things are 'safe' people don't get hurt? So, it shouldn't come as a surprise that the focus for many organisations is to have as little incidents and accidents as possible. But does the absence of incidents indicate the presence of safety? What if we use a medical analogy to test the logic of this statement? Does the absence of illness indicate the presence of health?

For these organisations, whether clearly stated in their strategy or not, safety is considered to be "a state where as few things as possible go wrong" (Hollnagel et al., 2015) and one way to measure this is to count the number of incidents or accidents over a given period of time. Under this model, perceptions of 'safety' develop when incident numbers drop or when long periods of time pass without an accident. At face value, this reasoning makes some sense, but in practice, it can actually become counterproductive to the organisations' goal of reducing injuries. This is because when things do go wrong, incident investigations usually identify employees as liabilities (often implicitly) due to the way they have interacted with a certain system, with training, with procedures or with technology. Blame creeps in very quickly and opportunities to learn are lost. People caused people to get hurt. People are the problem. People are hazardous. People need to be managed better. This message echoes loudly through the workforce and over time, people stop reporting incidents and near misses. Eventually (or sometimes, quite quickly), safety is viewed as just another disengaging management function.

In traditional safety approaches, risk assessments and incident investigations feature heavily. They are the stars of the safety production, with incidents taking centre stage. Risk assessments are used to try to identify the hazards that could contribute to an incident, and investigations are used to try to identify why an incident occurred. The core assumptions of traditional safety approaches make workplace safety reactive, prompting action when identified risks are above a certain threshold, or when an incident investigation identifies a 'corrective action'. The trouble with this is that most of the time, risk assessments and incident investigations infer human error as causal factors, pushing learning opportunities related to system improvements

DOI: 10.1201/9781032679525-3

or deficient defences farther away, while implicitly telling workers to stop openly sharing information that could help prevent future incidents.

Another potential trap of these approaches is that workplace health and safety can become additive very quickly. Risk treatments can result in more safety procedures or checklists, and corrective actions can lead to an overdose in safety toolbox meetings, death by training, or more punitive details added to JSAs. And before they know it, organisations find themselves carrying bloated safety management systems that the workers cannot relate to, despite the relentless procedural pushdown efforts from management and the safety team.

The fundamental flaw of traditional safety approaches is that they fail to examine why most of the time, human performance goes to plan (Conklin, 2019). And because the reactive nature of traditional safety approaches prompts action when things go wrong, learning opportunities are significantly diminished. Things go wrong far less often than they go right. So, doesn't it make sense to maximise our learning by investigating why things go right, rather than constraining our learning to when things occasionally go wrong?

Another feature of many traditional safety approaches is behaviourism; demonstrated through behavioural-based safety (BBS) programs. BBS adopts the principles of operant conditioning; a type of learning in which people are more likely to repeat behaviours that have been rewarded and less likely to repeat behaviours that have been punished. It can be quite effective on young children, and dogs (of all ages). Not so effective on employees. From the perspective of the worker, however, BBS reinforces that workers' behaviour is the cause of most work-related injuries and illnesses. This reinforcement can take the form of plainly expressed, explicit employer messaging and attitudes around employee behaviour, or implicitly implied through management responses to incidents or failure, safety programs, and organisational culture. Think about it for a moment. The title says it all. Behaviour-based safety; the behaviours of the workers determine if a state of safety is achieved or not. Their attitudes, motivations, biases, lived experiences and perceptions are not considered. These things are not observable. They don't form part of BBS observations. Individual factors, social factors, organisational factors, psychological factors and work situational factors collectively, and individually impact how we behave, yet these are overlooked by short-sighted safety programs such as BBS.

Having worked on several mine sites and civil construction projects as a blue-collar worker where BBS was the poster child for 'best practice' in safety, my direct experience on the other side of these programs was not positive. As a worker, being the subject of BBS observations was nerve-racking! Just knowing that a supervisor, line manager or safety officer was observing my 'safety behaviours' was enough to put me on edge. And mistakes often occur when people feel nervous or self-conscious. The point I am trying to make here is that many BBS programs interpret behaviour as a violation rather than being adaptive. How workers interact with organisational systems, or how systems impact decision-making and behaviours, or restrict autonomy, is rarely considered in BBS programs. This results in punitive consequences that rob management of any learning or insight into how workers navigate the variability of work. There are plenty of safety practitioners who still swear by BBS though, claiming that improvements in organisational and individual safety performance are a direct result of their BBS programs. However, I am yet to meet an advocate of BBS who has been able to convince me that the value of its utility and effectiveness has any positive bearing on the outcome of 'safety' – which is forever a temporary outcome mind you. This is a bias of mine and I am aware of it.

As a safety practitioner, I still find organisations are fixated on the behaviours of their employees. Whether it is driven from the human resources department, operational departments, or the safety department, a fair bit of effort goes into trying to understand and influence employee behaviours to meet organisational expectations. Codes of conduct, espoused company values, and safety rules ask for certain behaviours to be practiced and prohibit other behaviours. Everywhere you look, management, HR, operational departments, and safety teams alike are focussing on the behaviours of employees.

The trouble I see with this though is that trying to change behaviours by focussing on behaviours just isn't effective. You cannot and will not change behaviour by changing behaviour (Conklin, 2020). It's like focussing on injury frequency rates to try to reduce injury frequency rates, which is an absurd notion (despite this being a common approach to safety among executive teams and Boards). Behaviours are an expression of attitudes, values, beliefs, and conflicting needs – what people think and feel about things, people, situations, social norms, or organisational processes. What they believe in, what they value, and the priority they give to one thing (or person) over another. Trying to influence employee behaviours without appreciating what contributes to them is a losing battle for any safety professional or people leader. Forget using positive and negative reinforcements or rewards and punishments to create lasting behavioural change. The opportunity to influence behaviours comes by recognising that behaviours are an outcome or expression of attitudes, which are formed through beliefs, values, motives, biases, and cognitions. Focus on these things instead. Put another way, behaviours are not the problem, they are an expression of the problem (Lloyd, 2020).

Acknowledgement that behaviours are not the issue is a crucial realisation for the safety professional and a huge leap forward in beginning to understand, relate to, and connect with workers. Afterall, connection builds trust, and it is in trust where we can be shown the real issues faced by the workforce.

I do recognise that my own biases and theory perseverance around BBS make it quite difficult for me to be convinced otherwise, but I always welcome a healthy debate around whether behaviourism has a place in the current safety landscape. However, if BBS programs work for you, then that's great (maybe just talk candidly with the workforce to see if it works for them too!). I'm getting ahead of myself. I'll talk a little more about behaviourism in Chapter 5.

I know I am firing a few shots across the bow at traditional safety approaches, but don't get me wrong – there is value to be found in these models. Traditional safety practices have evolved over decades, and it would be foolish to discard these practices completely. There is value in conducting risk assessments, there is value in facilitating incident investigations (when conducted thoughtfully and with care), and there is value in having robust safety procedures that help employees conduct their work in safety-conscious ways. It is the core assumptions of traditional safety approaches that can trip organisations up. Their focus on failure, viewing people as hazards and their causality beliefs can be quite limiting. And we have these core assumptions to thank for statistically invalid safety measures such as total recordable injury frequency rates (Hallowell et al., 2020). Lots of traditional safety practices can be incredibly beneficial to organisations if they are to challenge these core assumptions with a human-centric lens. You don't need to change the game perse, you just need to change the way you approach it.

There are plenty of traditional safety aspects that I adopt in my safety approach that help drive systematic risk reduction for the organisations I work with. Safety-I did not screw up the safety industry like many 'new view' zealots will have you believe. Lots of things about traditional safety still make sense in today's world – but at the same time, lots of things about it don't. Sure, there are aspects of traditional safety that are bureaucratic and dysfunctional, and there are elements that can actually dimmish the safety of work, but my advice to any safety practitioner would be to seek to understand the elements of traditional safety approaches that are relevant to the needs of the organisation you work with and add them to your toolkit. There is some really good stuff in these traditional safety approaches, so go and look for it. Find what works for you, chew the meat and spit out the bones. Become informed enough to adopt the opportunities it offers, aware enough to avoid its traps, humble enough to explore your own core assumptions around safety, and willing enough to rebuild them on a foundation of humanism.

Chapter Summary

- Traditional safety models typically position safety as being reactive, with perceptions of 'safety' developing when incident numbers drop or when long periods of time pass without an accident.
- Traditional safety approaches run the risk of becoming counterproductive to organisations' goals of reducing injuries because when things do go wrong, incident investigations bring forth blame and opportunities to learn are lost.

- A blame environment drives incident and near-miss reporting underground due to low psychological safety.
- A potential trap of traditional safety approaches is that workplace health and safety can become additive very quickly, resulting in safety clutter.
- The fundamental flaw of traditional safety approaches is that they fail to examine why most of the time, human performance goes to plan.
- Using positive and negative reinforcements, or rewards and punishments to create lasting behavioural change does not work.
- Behaviours are an expression of attitudes, values, beliefs, and conflicting needs. Behaviours are not changed by changing behaviours.
- Traditional safety practices have evolved over decades, and it would be foolish to discard these practices completely. Identify the elements of traditional safety models that will benefit safety in the context of your organisation.

References

Conklin, T. (2019). *The 5 principles of human performance. A contemporary update of the building blocks of human performance for the new view of safety.* Amazon Digital Services LLC - KDP Print US.

Conklin, T. (2020). *Pre-accident investigations. An introduction to organisational safety.* The Podcast. Safety Moment - Changing Behavior by Changing Behavior. https://preaccidentpodcast.podbean.com/e/safety-moment-you-cant-change-behavior-by-changing-behavior/

Hallowell, M., Quashne, M., Salas, R., Jones, M., MacLean, B., and Quinn E. (2020). *The Statistical Invalidity of TRIR as a Measure of Safety Performance.* Published by The Construction Safety Research Alliance, Colorado.

Hollnagel, E., Wears, R.L., and Braithwaite, J. (2015). From Safety-I to Safety-II: A White Paper. The Resilient Health Care Net: Published Simultaneously by the University of Southern Denmark, University of Florida, USA, and Macquarie University, Australia.

Lloyd, C. (2020). *Next generation safety leadership: From compliance to care* (1st ed.). CRC Press. https://doi.org/10.1201/9781003051978

3

Progressive Safety Approaches

Let's revisit the primary objective of workplace safety again; to prevent employees from getting hurt. But this time, instead of focussing on why things go wrong, why don't we focus on why things go right, rejecting the notion that incidents are usually caused by human error and instead looking at how problems with organisational systems can affect human factors, viewing safety as something that emerges from these systems and from normal work. This is how progressive safety approaches such as Safety-II, Safety Differently, and Human and Organisational Performance (HOP) approach workplace safety.

While traditional safety approaches use rules, standards, and procedures to tackle this primary objective of not hurting workers, progressive safety approaches attempt to build sustained resilience by focussing on the human capacity to work safely in varying conditions. That's a bit of a mouthful I know, but if you look into any of these progressive approaches, you will quickly find they can be broken down into four or five principles that are pretty easy to understand.

One way to contrast traditional safety approaches with progressive ones is to take the examples used earlier, being risk assessments and incident investigations. Traditional risk assessments ask questions related to the job steps, the risks associated with those steps, the measures in place to mitigate those risks, the training, skills, and expertise of the workers, and whether the right tools and PPE are being used. They focus on hazards and more often than not, they are developed by safety teams or management with limited input from the workforce.

Progressive safety risk assessments ask questions related to elements of the job that are difficult, where mistakes can be easily made, where ambiguity exists in procedures, where work conditions could present adherence challenges, and where things need to change. These risk assessments focus on "error traps" rather than hazards, and answering these questions relies on substantial input from the workforce. And who says that risk assessments need to be written documents anyway? Safety legislation doesn't specify this. Let's get creative by trying to think of less coercive, and more engaging ways to conduct risk assessments and other safety activities with the workforce (rather than doing safety to the workforce).

As for incident investigations, under the traditional approach, the intent of these exercises is to attempt to identify the root causes of what went wrong

DOI: 10.1201/9781032679525-4

so changes can be made to help prevent similar incidents occurring in the future. Under progressive safety models, the intent of incident investigations is to attempt to understand why things usually go right, to help explain why occasionally, things go wrong. The HOP philosophy goes one step further, encouraging organisations to conduct pre-accident investigations on critical tasks that have not yet resulted in an incident.

Whilst there are subtle differences among these progressive safety approaches, one thing they all have in common is that they don't view people as hazards or problems to control. The workforce is acknowledged for the skills and expertise they have developed over time, and this is harnessed to help improve the safety of work. Unlike BBS models, employee behaviour is not viewed as a problem – it is viewed as an expression of the problem (Lloyd, 2020), which leads to intervening in work conditions rather than intervening in employee behaviour. And when the focus is to improve work conditions and how employees interact with organisational systems, our people become an integral asset in identifying solutions. People are not the problem; they are the solution. Prescription is swapped out with participation and the inferred blame that accompanies incident investigations, BBS, and paternalistic safety programs is promptly extinguished by an authentic learning environment.

This philosophy of measuring safety by the presence of positive worker capacities rather than the absence of incidents and unwanted events makes a lot of sense to me. Think of how a professional football team is coached. The players are taught and developed to replicate the actions and strategies that are known to help the team win the match. The focus isn't on how *not* to make mistakes. Sure, that's a small part of it, but coaches model how past and present greats of the game play and encourage their players to replicate these successful playing styles. The coaches do not focus on the worst players and instruct everyone to simply not to be like them. That's not how to coach a team. There are plenty more opportunities to learn about what goes right and why in a game of football, rather than focussing on the handful of instances where things go wrong. It is the same for workplaces. Enhance the conditions and capacities that help things go right, and you're going to have more wins.

If you have picked up that I am much more complementary of progressive safety approaches than I am of traditional ones, you would be right. However, I do believe there are some things to look out for in these new views of workplace safety. Firstly, I want to acknowledge that moving away from a traditional safety approach can be a difficult step for Executive teams and Boards. It can be tricky to articulate the benefits in a way that doesn't scare them, and it takes some careful reassurance and influencing, followed by good change management. Sometimes, they might think of these progressive approaches as a fad or *bandwagory*, in which cases buzzwords such as

Safety-II or Safety Differently will do more harm than good, so my advice is to arrive at that conversation prepared (if you wish to have it).

Secondly, I would caution any safety professional against doing away with all elements of a traditional safety approach in place of a purely progressive one, because just like I pointed out earlier, there are elements of all safety models that add value, and there are elements to be weary of. Just as context drives behaviour at a worker level, organisational context should inform which elements of which models are best suited to the needs of your organisation. Again, chew the meat and spit out the bones. Build your own toolkit.

Thirdly, I don't like to use terms such as Safety-I and Safety-II, or 'the old view' and 'the new view'. I found it quite difficult avoiding them in writing this chapter, landing on 'traditional safety approaches' and 'progressive safety approaches,' which are not much better! But the reason I try to avoid these labels is that advocates of traditional safety approaches can feel admonished by labelling such as Safety-I and Safety-II, which come with inferences that their skills and expertise are outdated and no longer valued. This is certainly not true. Traditional safety approaches have been heavily researched with decades of data for advocates or opposers to draw from, however progressive approaches do not share this benefit, fuelling disapproving views across the industry from sceptics, as they chant in unison, "show me the proof!" It's a fair request. Where is the proof?

I acknowledge how difficult it is for senior management, Executives and Boards to deviate from a deep-rooted and unquestioned way of thinking about safety. I also acknowledge the organisational, social, and personal barriers that need to be overcome in order to embark on a progressive safety journey. So, why not drop the labels and just build your safety programs around your people and their needs? Humanising safety doesn't need to come with a new title or catch phrase. All it needs is a good understanding of what safety means in the context of each organisation, and the willingness to listen to, and accommodate the diverse perspectives of the workforce all the way up to the Board. And it is vital that as safety professionals and leaders, we practice self-awareness by recognising when our own biases, mindsets, or power imbalances present the temptation to take workplace safety down a path to satisfy our own needs rather than those of the collective for whom it is intended to serve. I've fallen into this trap before, and I'll probably fall into it many more times in the future. It's a tricky thing to be truly self-aware.

What I hope to show you throughout this book is that regardless of which safety models you prefer, or which elements of traditional or progressive safety resonate with you, if you approach your work with a human-centric lens, and apply the four steps described in future chapters, you are going to have more wins than losses.

Chapter Summary

- Acknowledging the workforce for the skills and expertise they have developed over time, and harnessing this, can significantly improve the safety of work.
- Contemporary or progressive approaches view safety as something that emerges from normal, everyday work and from the interaction between organisational systems and people.
- Progressive safety approaches attempt to build sustained resilience by focussing on the human capacity to work safely in varying conditions.
- Progressive safety risk assessments ask questions related to elements of the job that are difficult, where mistakes can be easily made, where ambiguity exists in procedures, where work conditions could present adherence challenges, and where things need to change.
- Under progressive safety models, the intent of incident investigations is to attempt to understand why things usually go right, to help explain why occasionally, things go wrong.
- Intervene in work conditions rather than intervening in employee behaviour.
- Moving away from a traditional safety approach can be a difficult step for Executive teams and Boards. Be prepared for this conversation.
- Organisational context should inform which elements of which safety models are best suited to the needs of your organisation.

MICRO-PROJECT #1

Defining Safety in Your Organisation

So, what is safety? It seems it's not such a simple question after all. Let's pull on that thread. Let's see if you can produce a workable definition with a bit of help from those within your organisation who are closest to it. Welcome to your first micro-project.

STEP 1

Try to gain management support to undertake the following exercise, outlining the basic steps and what you hope to achieve from it. Feel free to use some of the points I have raised in these past chapters around the need for organisations to define safety, and really try to plant the seed with management that this is the first step towards humanising safety – using real perspectives of the people who make up the organisation, to help inform how to progress workplace health and safety in the context of the work of the organisation.

STEP 2

Present the following questions to a healthy representation of the organisation. My suggestion is to include each member of the Executive team, a few people in middle management (like department heads or divisional managers), a group of team leaders, and a decent cohort of the frontline workforce. Let everyone know that this is an exercise aimed at helping your organisation understand what it needs to progress its safety journey and let them know you won't be personally identifying them in the responses they provide. And don't leave yourself and your safety team out either!

1. What is the definition of safety in our organisation?
2. What does good safety look like?
3. What does bad safety look like?

Ask each person to respond in writing and give them all a couple of weeks to do so. This first step is all about collating different perspectives before having the broader conversation. Once you have received your responses, make sure no one is identifiable and arrange them in random order on a document or PowerPoint presentation. Keep a record of responses by group for step 3.

STEP 3

Schedule a meeting with senior management (those ultimately responsible for workplace safety) to go through all the responses provided. The objective of this meeting will be for senior management to agree on a definition for safety. During this session you can restate the intent of the exercise, summarise the process and groups of staff involved, and then share the responses to prompt some candid discussion. If there were some notable differences in responses between groups, you might want to highlight these too and prompt senior management to share why they think perceptions might differ depending on where in the organisation people work. There will likely be some really useful insights shared during this meeting so you might want to take some notes. I've found these exercises hugely beneficial in helping identify who my most powerful safety allies in the organisation are.

STEP 4

Once senior management has agreed on its definition of safety, make sure you share it with all those who provided responses to the questions. This feedback loop is an important step in building trust and showing that their voices have been heard. Well done! You now have your definition of safety from which you can start building your strategy.

Reference

Lloyd, C. (2020). *Next generation safety leadership: From compliance to care* (1st ed.). CRC Press. https://doi.org/10.1201/9781003051978

Part 2

The Psychology of Workplace Safety

4

Embracing the Human Condition and Its Contribution to the Safety of Work

I am really excited to write this second part of the book, simply because I am fascinated by the psychology of human behaviour in terms of workplace health and safety, and I want to share what I have learnt with as many other safety professionals as I can. I won't be able to cover nearly as much as I want to in these next few chapters, but hopefully, I can highlight some links between psychology and safety, and hopefully by touching on some of those links, it might pique your interest just enough for you to delve deeper into them, or discover other links, to not only make you a better safety professional or leader but also to help you realise how your influence impacts the humans in your organisation – for better or worse. Being a safety professional is a privilege. These are highly influential roles, and there is significant responsibility that comes with this privilege.

I think I should start with a bit of a disclaimer for the next few chapters. Just to share with you what to expect as you make your way through this part of the book. I have tried not to write the next few chapters too academically, but given the lens is psychology, it is difficult to avoid entirely. For some readers, this might be right up your alley, and for others, it might not be. That's cool – we're all different. Despite this, I really encourage you to approach the next few chapters with a curious mindset because I passionately believe that some of these concepts contribute to the foundation of a humanised safety approach. So, with that said, let's get into it!

What is the human condition and why is it relevant to workplace health and safety? This is a big question, but it seems like a logical place to start when talking about human-centric safety.

The human condition is the collection of emotions, states of being, shared experiences and key life events that are common to all of us, regardless of our backgrounds, cultures, preferences, or race (Practical Psychology, 2020). Happiness, sadness, empathy, curiosity, trust, fear, loneliness, belonging, learning, aspiration, conflict, motivation, and purpose – all these emotions and human attributes turn up for us in different ways and at different stages across our lives. The human condition is complex – and I'm not equipped to explain it in detail, but I understand the basics and I also understand the rabbit holes a full-blown explanation attempt could send us down. Even thinking about the human condition could be considered part of the human condition! Yikes!

DOI: 10.1201/9781032679525-6

As safety professionals, I think we should all understand the rudimentary elements of the human condition and how they relate to workplace health and safety. It is something we all have in common yet at the same time, something so uniquely individual to each of us. The main point to note here is that as safety professionals, we need to understand it, embrace it, and factor it into our safety programs if they are to be successful. Effective safety management is deeply rooted in the human condition – the personalities, emotions, lived experiences, and characteristics of the people we work with are absolutely critical in understanding workplace behaviour and its contribution to the safety of work. Understanding who our people are, what they are passionate about, what they are disinterested in, what motivates them, what demotivates them, how their thoughts inform their emotions, and how their emotions determine their behaviours. Understanding your people is the best tool any safety practitioner can have because effective safety management relies on connection, and connecting with your people is not possible without first understanding them and then building trust.

Let's touch on trust for a moment. Can you think of a past (or present) manager who adopted a micro-management approach, standing over your shoulder and giving you detailed instructions on how to perform your work, with little appreciation for the skills and expertise you have brought to the table? Can you also recall a manager who led with trust, allowing you the autonomy to do the work you were hired for as an expert in your field, making scaffolding available for the times you require support? To the employee, these management approaches feel very different. Being told what to do signals danger to the central nervous system, whereas being trusted signals safety. Think about how you might behave when you feel fearful, compared with how you might behave when you feel trusted. In which state do you think accidents are more likely to occur?

Trust and fear are on opposite ends of a continuum, and elements of the human condition that are critical to understanding how people engage with workplace safety programs. Let's take an example. A real example.

Jason was a Safety Manager for a reputable engineering and construction company in Melbourne, Australia. He was well respected by management with a standing in the organisation that he described as "being appropriately feared by the workforce". Over his years of experience as a safety manager, Jason had learnt that management responded well to high levels of safety compliance, and Jason's method of maintaining or increasing compliance was to apply an authoritarian approach to safety, underpinned by what he had learnt about behaviour-based safety in a three-day conference five or six years earlier. Jason would reward good safety behaviours with verbal recognition, a pat on the back, and at times, a nomination for the annual safety award under the category of 'demonstrating safe behaviours.' However, if Jason was told of, or observed a worker behaving in a way that was deemed by him as being unsafe, he would punish that behaviour with verbal reprimand, reinduction or additional safety training, and at times, redeployment to a less risky (less meaningful) task until the worker's bank of 'safe behaviours' was built up again.

For the most part, this approach worked for Jason. The workers knew exactly what types of behaviours they would be rewarded for, and what types of behaviours they would be punished for (if they were caught). The challenge this approach presented was that the workforce operated from a state of fear, and safety on Jason's projects was determined purely by observable behaviours. The workers did not trust Jason – and therefore did not trust management's approach to safety, and their fear of reprimand meant a rigid application of safety procedures, despite the impracticality of those procedures and the variable nature of the work they performed. There was no oxygen around safety. It was a tangled knot of dogmatic processes. People weren't having open safety conversations, management was blind to the real safety challenges experienced onsite, workers were too nervous to raise safety concerns, and near misses and incidents were only reported if someone saw them occur. But compliance levels were high!

Leading with fear may result in temporary gains in adherence to safety processes but it stifles safety climates, masking the actual safety experiences of the workforce and pushing management even further from their people.

There was a great ending to this story though. The joint venture between the client and the construction company Jason worked for, ended up engaging the refreshingly talented Clive Lloyd to run his *Care Factor Program* with the leadership teams across the project. After participating in the program, Jason was self-aware enough to humble himself, review his approach, and commence the very difficult task of rebuilding trust with a workforce that feared him. Last time I checked in with Jason, he had started working with a new construction company and was experiencing great success leading with trust rather than fear.

While this story highlights the importance of how trust and fear can play into safety attitudes, it is equally as important for the safety professional (and leader) to consider how other emotions, lived experiences and key life events can interact with the safety of work. Humans are emotional beings, and our emotions play a key role in determining how we behave at work. We all make adjustments in our personal lives when we experience strong emotions. Work should be no different. Our emotions prepare us for behaviour, orchestrating our perception, attention, learning, motivational, and behavioural decision-making systems (Cosmides & Tooby, 2000; Tooby & Cosmides, 2008). Emotions and safety behaviours play in the same sandpit. The interplay between our emotions, our experiences and our work are fundamental elements of the human condition and how it contributes to the safety of work. It deserves a lot more consideration from employers and safety professionals than it currently gets, and the pathway starts with getting to know our people.

Chapter Summary

- The human condition is the collection of emotions, states of being, shared experiences and key life events that are common to all of us, regardless of our backgrounds, cultures, preferences, or race (Practical Psychology, 2020).
- Effective safety management is deeply rooted in the human condition – the personalities, emotions, lived experiences, and characteristics of the people we work with are critical in understanding workplace behaviour and its contribution to the safety of work.
- Understanding your people is the best tool any safety practitioner can have because effective safety management relies on connection.
- Think about how you might behave when you feel fearful, compared with how you might behave when you feel trusted. In which state do you think accidents are more likely to occur?
- Humans are emotional beings, and our emotions play a key role in determining how we behave at work.

References

Cosmides, L., & Tooby, J. (2000). Evolutionary psychology and the emotions. In M. Lewis and J. M. Haviland-Jones (Eds.), *Handbook of emotions* (2nd ed., pp. 91–115). Guilford Press.

Practical Psychology. (2020, June). *The Human Condition (Definition + Explanation).* Retrieved from https://practicalpie.com/the-human-condition/.

Tooby, J., & Cosmides, L. (2008). The evolutionary psychology of the emotions and their relationship to internal regulatory variables. In M. Lewis, J. M. Haviland-Jones, & L. Feldman Barrett (Eds.), *Handbook of emotions* (3rd ed., pp. 114–137). The Guilford Press.

5

Relating the Five Major Psychology Perspectives to Workplace Safety

When people refer to perspectives in psychology, they are talking about the different approaches or theoretical frameworks adopted by psychologists to study human behaviour (Mcleod, 2023). There are five major psychology perspectives, each involving certain assumptions about how people think, feel, and behave. These are the psychodynamic, biological, behavioural, cognitive, and humanistic perspectives, and just like safety approaches, each perspective has its strengths and weaknesses – that is to say, there are no right or wrong perspectives. I am not going to labour on these too much. I just want to shine a light on the links that can be drawn between the five major psychology perspectives and workplace health and safety. So, if you have the time and the interest to explore these linkages, I really encourage you to do so, but for now, here is a quick crash course that only really scratches the surface. You might need to grab a coffee for this next part, as it might be a bit dry for some readers! -pause to boil the kettle-

The psychodynamic perspective is the birthchild of Sigmund Freud, the founder of psychoanalysis, who believed that the unconscious mind together with childhood experiences determine our behaviour and that humans have little free will (Bornstein, 2024). Although this perspective was developed way back in the 1880s, most of us still use psychodynamic concepts in our lives and some of these remain relevant to the field of workplace safety. For example, operating from the subconscious mind (reduces situational awareness), denial (inhibits learning), and transference (projecting feelings onto others can increase psychosocial and physical risks in the workplace). These are all psychoanalytic concepts that are relevant to helping understand worker behaviour, and understanding the motives and emotional responses that influence worker behaviour.

One of the contentious, core assumptions of psychodynamic theory is that there is nothing random about our thoughts, feelings, emotional responses, motives, or behaviours (Bornstein, 2024). That's not very exciting, is it? This theory proposes that none of the elements of our mental lives happen by chance – not even the most random thought such as where to sit during a safety toolbox meeting. Instead, it is argued that every thought, feeling, motive and expressed behaviour results from biological or psychological influences (Elliott, 2002) and therefore, we have little, if any, free will. What do you think about this claim? Do you agree that people do not have free will

DOI: 10.1201/9781032679525-7

and that our thoughts, feelings, and behaviours are never random? Or do you rebuke this claim? How does your position impact how best to approach workplace safety, if at all?

What about the risks associated with operating in autopilot mode? Do the safety practices within your organisation consider that people default to operating from their subconscious mind at work (autopilot) due to the simple fact that they are human? Is 'autopilot' recorded as a risk in your risk register? Should it be?

Regardless of where you stand on these things, I hope it makes you think a little deeper about the connection between our mental lives at work, the concept of free will, and safety as an outcome.

The biological perspective views our thoughts, feelings, and behaviours as being caused by biological factors, such as our brain structures, hormones, the immune system, nervous system, and genetics. Looking at workplace safety through this perspective, things become interesting when you consider the biological explanation for our fight, flight, freeze response in the face of stress – including our stress response to workplace danger. How stress responses differ from person to person and what this might mean in the event of an emergency, or when on-the-job tasks become suddenly dangerous. Consider also, the impeded decision-making capacity of a supervisor or manager when faced with high stress due to workload, and the safety implications this can result in downstream for the workforce.

Another potential area to explore in the context of workplace safety from the biological perspective is the influence neurotransmitters have on our psychological functions. These chemical messengers help our brains and nervous systems communicate, influencing our cognition, moods, and emotions. Dopamine (motivation), serotonin (mood), glutamate (memory), noradrenaline (concentration), and acetylcholine (learning). These neurotransmitters are all relevant to workplace safety, right? The safety profession is deeply invested in *motivating* employees to work safely, committing safety messaging to *memory*, maintaining task *concentration*, and helping employees to continuously *learn*. Understanding the biological basis of human behaviour is an advantage to any safety professional and for this reason, studying the biological perspective of psychology can be a good investment.

The behavioural perspective is contentious in a safety context and a bit of a departure from the other four perspectives in that it views behaviour as the direct result of environmental factors. In fact, it goes so far as to state that the environment determines all behaviour and, similarly to the psychodynamic perspective, that people do not have free will. Most behavioural studies have involved animals, and some critics of behaviourism argue that animal studies cannot simply be generalised to humans because human behaviour is much more complex (Araiba, 2019). The behavioural perspective proposes that people learn from the environment according to two main processes: classical conditioning and operant conditioning (Mcleod, 2023). Learning by association where two stimuli are repeatedly paired, is known as classical conditioning. Think about how you might start to salivate when you see an ad on TV about a food you love, then after some time, you find yourself salivating whenever you simply spot the logo for that food or hear the jingle. Or how a song might take you back to a very specific memory, triggering an emotional response like feeling carefree or in love. These are examples of classical conditioning and classical conditioning is adopted in several safety campaigns – particularly using fear as the conditioned response in attempts to discourage unsafe behaviours. I'm sure you can think of a few examples.

Operant conditioning uses rewards and punishment to modify behaviour and is the basis of behavioural-based safety (BBS) programs. Desired 'safe' behaviours are rewarded, and undesired 'unsafe' behaviours are punished. BBS programs can send the message to the workforce that people are the problem and therefore, their behaviour must be corrected if the organisation is to improve its safety performance. For advocates of BBS, I encourage you to do some research on the behavioural perspective of psychology, attempting to search for any scientific evidence showing that BBS programs actually reduce the likelihood of injuries, and then ask yourself if you still think it is possible to punish and incentivise your way to safety.

In a world where safety clutter reigns, **the cognitive perspective** is perhaps the easiest to draw synergies with workplace safety. This psychology perspective is the study of how people think, remember, and learn. Memory, perception, language, attention, decision-making, and problem-solving are all areas of study in cognitive psychology. Think of the human mind as a computer. Of all the billions of pieces of information available to our brains throughout the course of a day, how do we know what to give our limited attention to? Studying how people process, filter, encode, store, and retrieve information should be of great interest to every safety professional. Especially when you consider the ridiculous amount of safety information, we expect workers to remember and retrieve at the drop of a hat. Safety inductions, training material, policies, procedures, job safety analysis (JSA), safe work method statements (SWMS), site safety rules, emergency response procedures, toolbox talks – you name it! It is completely unreasonable (and frankly impossible) to expect workers to remember all this information and be able to retrieve it when job conditions change. Safety professionals can start to reduce this safety clutter by understanding the scientifically proven limitations of the human brain that cognitive psychology research offers and begin to brainstorm more effective ways to communicate (relevant) safety information. If you think about all the stimuli assaulting the senses of your workers out there in the field, you might come to appreciate how difficult it is for them to focus on the safety of their work. We need to cut them a break.

Behaviour is informed by human cognition, so researching the cognitive perspective can help us understand why people exhibit safe and unsafe behaviours. Makes sense, right? Liu et al. (2023) defined safety cognition as "the process of obtaining and processing safety-related or hazard-related information and implementing decisions". Just pause for a minute to consider how difficult it might be for an employee to recall and process relevant safety information among the sea of safety noise and safety clutter they face in their role. Does all this safety junk help them make decisions or does it impede their ability to make safer decisions? I like to use an analogy about choosing what to watch on Netflix. If there are loads and loads of options, it can take me ages to pick a movie or show. But, if there are only a handful of options, I make my choice must faster and I rarely regret it.

I think the safety profession still largely blames individuals for cognitive errors, failing to accept that the information processing capacity of the human brain is actually quite limited. There are plenty of studies on cognitive failure that attribute incidents or unwanted outcomes to cognitive-based errors, but this viewpoint often assigns blame on the individual instead of addressing how cognitive overload can detrimentally impact one's decision-making ability, as is often the case with safety noise or safety clutter. I think the safety profession needs to realise that force feeding our workers more safety procedures, more safety training, and more safety toolboxes can actually introduce hazards, by overloading their cognitive limitations and making it increasingly difficult for them to make decisions. What do you think?

Well-being, free will, self-actualisation, and self-efficacy. If these resonate with you, then you are likely going to appreciate what **the humanistic perspective** of psychology has to offer. This perspective is built around a couple of key assumptions, being that human beings are inherently good, they flourish under the right conditions, people have free will, and we are all responsible for our own self-actualisation (Krems et al., 2017). Humanistic psychology emerged in the 60s as a divergence from the deterministic approaches of psychoanalysis and behaviourism, offering a new philosophy built around the concept of the whole person (*The Journal of Humanistic Psychology*, 2023). Now, let's see if we can relate this perspective to workplace safety.

In the safety profession, do we assume people are inherently good? Do people thrive under the right working conditions? Do employees have free will with the ability to behave at their own discretion? Are workers responsible (at least to some degree) for their own self-actualisation? I'm fairly confident you answered "yes" to most, if not all of these questions – or at least I hope you did!

Employees want to do the right thing and they want to keep themselves and others safe. They, like the rest of us, are inherently good. The safety profession must recognise this. And we know that under the right work conditions, employees thrive and can conduct their work safely, right? Therefore, shouldn't the focus be on helping create the right work conditions and recognising that our people do have free will, rather than adopting a deterministic, behavioural view of the person that puts workers on the hook and removes safety accountability from management? Some of the most successful safety programs I have been involved in have focussed on building employee self-efficacy and self-actualisation, with well-being being a central pillar. Thanks humanistic psychology!

So, there you have it. I have just (very roughly) taken you through the five major psychology perspectives, drawing connections between each perspective and workplace safety. I tried to keep it brief and I did try hard to stop myself from rambling on too much. You may have noticed that I did not select one perspective and discard the others (although I was not very complimentary of behaviourism). I try to adopt an eclectic approach when applying these perspectives to workplace safety, and it should be no different when it comes to safety models. Safety-I, safety-II, safety differently, HOP, human factors, 'the new view'. You may prefer one over another, but don't let your preferences and biases blind you from the opportunities that reside in the others – even if those opportunities come in the form of showing you what not to do! Build your own eclectic approach to workplace safety, being sure to put the workforce at the centre, just like a delicious pickle in a burger!

Chapter Summary

- There are five major psychological perspectives, each involving certain assumptions about how people think, feel, and behave.
- Examples of the psychodynamic perspective in safety include operating from the subconscious mind, denial, and transference.
- The biological perspective views our thoughts, feelings and behaviours as being caused by biological factors, such as our brain structures, hormones, the immune system, nervous system, and genetics.
- Understanding how neurotransmitters can affect our psychological functions has relevance in workplace safety. Dopamine (motivation), serotonin (mood), glutamate (memory), noradrenaline (concentration), and acetylcholine (learning) all contribute to states of safety and well-being.
- The behavioural perspective views behaviour as the direct result of environmental factors and BBS programs are built from this perspective.
- BBS uses operant conditioning in an attempt to reward and punish a pathway to safety. BBS programs can send the message to the workforce that people are the problem and therefore, their behaviour must be corrected if the organisation is to improve its safety performance.
- The cognitive perspective is the study of how people think, remember, and learn. Memory, perception, language, attention, decision making, and problem-solving are all areas of study under cognitive psychology, relevant to workplace safety.

- The information processing capacity of the human brain is actually quite limited.
- The humanistic perspective posits that people are inherently good, they flourish under the right conditions, they have free will, and we are all responsible for our own self-actualisation.
- Employees want to do the right thing and they want to keep themselves and others safe. They, like the rest of us, are innately good.

References

Araiba, S. (2019). Current diversification of behaviorism. *Perspectives on Behavior Science, 43*(1), 157–175. https://doi.org/10.1007/s40614-019-00207-0

Bornstein, R. (2024). The psychodynamic perspective. In R. Biswas-Diener and E. Diener (Eds.), *Noba textbook series: Psychology.* DEF Publishers. Retrieved from https://noba.to/zdemy2cv

Elliott, A. (2002). *Psychoanalytic theory: An introduction*. Duke University Press.

Krems, J. A., Kenrick, D. T., and Neel, R. (2017). Individual perceptions of self-actualization: What functional motives are linked to fulfilling one's full potential? *Personality & Social Psychology Bulletin, 43*(9), 1337–1352. https://doi.org/10.1177/0146167217713191

Liu, Y., Ye, G., Xiang, Q., Yang, J., Goh, Y. M., & Gan, L. (2023). Antecedents of construction workers' safety cognition: A systematic review. *Safety Science, 157*, 105923

Mcleod, S. (2023). *Theoretical perspectives of psychology (Psychological approaches)*. Published by Simply Psychology.

The Journal of Humanistic Psychology. Gale General OneFile. link.gale.com/apps/pub/0DXJ/ITOF?u=ntu&sid=bookmark-ITOF. Accessed 7 December 2023.

6

Building a Healthier Relationship with Failure

Poke the system and it will reveal itself to you.

David Snowden

There are two key points I will attempt to make in this chapter. The first is that focussing on failures in a safety context can rob organisations of the numerous success stories that otherwise go untold. The second is that although it is well known that humans learn through failures, the negative stigma surrounding failure or mistakes can prevent them from being talked about, and therefore, paradoxically, we fail to learn from failures.

The unfortunate commonality among most organisations is that safety is all about measuring failures rather than successes. Injury rates make their way up to executive and Board reports, whilst tasks performed successfully and without incident don't get a mention. Safety is a measurement of failures rather than a measurement of defences. I get it though – finding errors in retrospect after an incident and reporting on them is much easier than finding defences that performed as intended, resulting in non-events. Consider the following statement in a Board safety report: "During the reporting period there were 7 incidents. These were investigated and contributing factors were addressed with mitigating measures put in place". There is nothing unusual about this statement. I've written similar statements in loads of safety reports. Now consider a statement such as this: "During the reporting period there weren't 180 incidents. These non-events were investigated and it was found that the defences put in place were effective and performed as intended". A statement like this probably wouldn't land well in a Board report. Reporting on non-events is difficult and isn't popular with executives and Directors who seem to be convinced that Director due diligence comes in the form of interrogating incident trends and audit findings, and little more. What went wrong and why it went wrong, rather than what went right and why it went right. There is more incentive in the eternal search for failures than there is in reviewing successful work. This is the unfortunate reality and management's responses to failure reverberate ominously under the boots of the workforce, reinforcing to them that failure is not received well and therefore mistakes are not tolerated. The messages surrounding safety being a measure of unwelcome failure erodes psychological safety,

DOI: 10.1201/9781032679525-8

as workers are reluctant to admit mistakes, are hesitant to report errors, and are fearful to report incidents and near misses. And as the reporting declines, the safety risk increases, stifling learning opportunities not only as a result of fewer errors and incidents being reported but also due to management's disinterest to learn from all the non-events; all those tasks that were executed to plan, with effective defences and without incident. Sadly, the opportunities to apply similar defences to other tasks are lost, as negativity bias continues to breed throughout the organisation, gnashing its teeth at any signs of positive recognition.

So, what can we do about this problem? How can we unpick this destructive mindset towards failure and safety? I think there are two areas for the safety professional to focus on. Firstly, we need to reframe failure as a learning opportunity (which I will talk about in a moment), and secondly, we need to shift towards measurements of safety that are based on the effectiveness of defences – the stuff that protects people from getting hurt. Because when you boil all this stuff down to a slow simmer, safety is more about the performance of defences than it is about the absence of incidents, right? Let's shine a spotlight on the defences that have been put in place to increase safety and find ways to integrate these into our measurement of workplace safety. Simple measures like new defences identified through pre-accident investigations, verifying the effectiveness of existing defences, or post-job reviews that identified new or improved defences. Exploring the vast library of resources around Critical Control Management will be a good starting point and will help draw focus on the performance of controls surrounding those high-consequence, low-likelihood risks that if they materialise, can really hurt people and the organisation. Weaving some lead indicators like these into your safety scorecards can start to subtly shift management's thinking from safety being the absence of injuries to safety being the presence (and performance) of defences. Think about the definition of safety that you came up with from the first micro-project. Does measuring failures (like incidents and audit non-conformances) provide you with sufficient evidence to determine if you are moving towards, or away from, your definition of safety? Or do you need to revisit your definition and have another go at it?

Now to the second key point, I wanted to land. The stigma around failure. The negative and unfair beliefs around errors and mistakes. Ask anyone if they make mistakes in life and if they answer honestly, their response will be a resounding "yes". Ask them again if they make mistakes at work, and you might find they are a little more reluctant to admit making mistakes in their profession. Why is this do you think? Is it simply a matter of self-preservation or ego, or is it the result of how we have been conditioned to view mistakes in a work context? We learn through our mistakes, right? At least that is what we teach to our children. So, shouldn't we be encouraging our people to recognise their mistakes and share them with the rest of us so we can all learn and improve? Can we change our relationship with failure by embracing it with a learning mindset, accepting that errors occur frequently in pretty much every aspect of life, including in work? People make mistakes all the time – it's part of the human condition. Errors also exist in systems, in designs, in practices, and in learning. Error is normal. Mistakes are normal. Failure is normal. Failures provide opportunities for organisational learning but only if it is framed this way and only if workers are encouraged and positively recognised for reporting errors, failures, or deviations from desired outcomes. Unfortunately, though, organisations approach failure reluctantly and respond to it unconstructively, and this is seen, heard, and felt by the workforce. Take the humble safety toolbox meeting as an example. Think about when a worker joins the safety toolbox meeting only to hear about the incidents that occurred over the course of the week and the resultant increase in safety training (like this toolbox) or other safety red tape. How do you think this impacts worker mindsets around human error and making mistakes? Does it send the message that reporting mistakes are welcomed for their rich learning opportunities? Or does it pull safety – something so inherently human – back into a polarising conversation around making fewer mistakes, having fewer incidents, and directing or inferring blame to a person or team?

What about the various standards of perfection inferred through organisational goals like "zero harm" or the intolerance for mistakes evident in highly prescribed safety procedures? What messages do these send to the workforce? There is no room for the human condition in these approaches. People make mistakes all the time, yet many workplace safety standards require perfectionism in the pursuit of poorly formed goals such as "zero harm". Anything less than zero is a failure – and failure is perceived as negative by workers, supervisors, team leaders, divisional managers, executives, and Directors alike. I think the mindset shift that is required here is to separate perfectionism from mastery. Those with mastery mindsets expect and prepare for errors, using them as opportunities to practice their 'error responses.' Perfectionists perceive failure when a mistake is made (like an injury in a "zero harm" organisation) and the window of learning is drastically narrowed as blame emerges in its place. Don't you think it makes sense to learn from our errors rather than sweep them under the carpet and pretend they never happened? Don't you agree it's time to rebuild our relationship with failure?

If you are interested in this line of thinking and concepts such as intelligent failure and learning from failure as a team, then I encourage you to read Amy Edmondson's book, *Right Kind of Wrong: The Science of Failing well.*

Chapter Summary

- Error is normal. Mistakes are normal. Failure is normal.
- Focussing on failures in a safety context can rob organisations of the numerous success stories that otherwise go untold.
- Although it is well known that humans learn through failures, the negative stigma surrounding failure or mistakes can prevent them from being talked about, and therefore, paradoxically, we fail to learn from failures.
- The unfortunate commonality among most organisations is that safety is all about measuring failures rather than successes.
- Any messages surrounding safety being a measure of unwelcome failure can erode psychological safety, as workers are reluctant to admit mistakes and are hesitant to report errors.
- Trust can be built by reframing failure as a learning opportunity and shifting towards measurements of safety that are based on the effectiveness of defences.
- Organisations can change their relationship with failure by embracing it with a learning mindset and accepting that errors occur frequently in pretty much every aspect of life, including in work.
- Goals like "zero harm" ask for perfectionism from workers in a world where perfectionism doesn't exist. Try to focus on mastery rather than perfectionism.
- Poke the system through 'safe to fail experiments' and it will reveal itself to you. It will either bite your finger off or reward you.

7

Safety as a Social Construct

Safety management systems, hazard identification, risk assessment, safety culture, and incident prevention are commonly considered some of the deterministic elements of safety in the workplace. Most organisations have metrics in place that report on these organisational conditions and you do not have to look far to find rigid models for each of these that promise to deliver safety outcomes to any organisation susceptible to the lure of fear-mongering sales campaigns. But do these often decontextualised, prescriptive, top-down models really work? Is workplace safety as tangible and quantifiable as mainstream safety wills us to believe? Is it an objective condition to be discovered by employees given the right set of organisational and environmental conditions? Is it something that can be proven through safety audits? Or, by contrast, is safety more of a social construct created through interactions, perceptions, and social norms, where employees are active agents in the creation of their safety knowledge? Perhaps it is a bit from both camps. What do you think?

Regardless, I am of the view that building a healthy understanding of the social construction of workplace safety benefits not just safety professionals and the stakeholders they seek to serve, but also the safety industry at large, given a social constructionist perspective of safety challenges many individualistic, mechanistic approaches that are stifling the profession and frankly, in many instances, doing more harm than good. To view safety as a social construct is to recognise that people develop their knowledge of safety in a social context, and what we perceive to be 'safe' depends on shared assumptions and experiences rather than 'safety' being an objective reality. And because shared assumptions change over time, our perceptions of safety will also change over time. Whilst this makes defining safety more difficult (recall micro-project #1), it opens the door to the fascinating realm of social psychology which I believe should be a staple in every safety professionals' learning.

The insights that can be gained through exploring the basics of social psychology – the study of how people affect and are affected by others, focussing especially on the power of situations (Baumeister, 2020) – is especially important in the context of workplace safety. Anyone interested in human behaviour (such as safety professionals and people leaders) will benefit from studying social psychology in an organisational context because understanding that behaviour is influenced by emotions, social processes, sociocultural factors, social comparison, and the presence and behaviours

DOI: 10.1201/9781032679525-9

of others, can not only help us understand each other, it also helps shift our perspective on workplace safety to a more humanistic space.

There are so many learning and growth opportunities for the safety professional in social psychology. However, for reasons that escape me, social psychology fundamentals do not form part of the syllabus in workplace health and safety curriculum. Social cognition, attitudes and beliefs, emotion, prosocial behaviour, antisocial behaviour, group formation, and social influence and persuasion. These all belong in the safety professional's toolkit! Let's take social influence techniques for example. Whether it's the disrupt-then-reframe technique, labelling techniques (assigning labels to people and then requesting behaviours consistent with the label), using the scarcity principle to gain buy-in to safety initiatives (rare opportunities are viewed as more valuable), or being aware of "advertisement wear-out" which occurs quite a bit in the world of safety and results in employees becoming inattentive or irritated after hearing the same message too many times. Knowing when and how to spot (or use) these influencing techniques can help safety professionals maximise their standing in an organisation as a key influencer of employees and management, or acting as an influential conduit between employees and management. Knowing when to draw attention to strong arguments, or when to disrupt attention away from unconvincing arguments, is a skill largely undervalued in most organisations, yet it can yield significant results if used correctly. A caveat here though – influence and persuasion techniques should only be used with good intent, for the benefit of workplace safety and not for self-serving purposes.

And what about emotions? Our behaviours are a product of our emotions, so surely emotion and affect should be taught in workplace safety curriculum, right? One of the key assumptions of social psychology is that our inner processes serve interpersonal functions. Our emotions help us maintain relationships and connection with others, and stronger connections and cohesiveness within a work team result in better decisions, increased efficiency, and greater employee satisfaction (Lynch, 2018). We would be mad not to learn about emotions! I think I am wandering off topic a little. Let's get back on track…

Workplace safety *is* socially constructed. At least to some extent. Safety is an ever-evolving social negotiation within every organisation, bringing together factors such as risk perception, the normalisation of unsafe work, the desensitisation of risk, and the politics of accounting for safety which often results in the manipulation of statistics (Turner and Gray, 2009). In my experience, these factors are notable influencers of organisational safety climates. Acknowledging them, and talking openly about them with management, can help bring safer workplaces to the foreground by facilitating richer, more honest conversations about what is making the attainment of safety goals difficult to achieve. Even starting a conversation on what leads to the normalisation of unsafe work can deliver some pretty powerful insights, based on the experiences of the humans working in your organisation. Unlike safety audits that more often than not, focus on documentation and surface compliance (and risk sparking an organisational *auditism* epidemic), having an open conversation about the normalisation of unsafe work can uncover the real issues faced by those who are closest to the work. In a study by Hutchinson et al. (2024), it was found that safety audits "exhibited a comprehensive shallowness, delving excessively into minor system details and paperwork rather than addressing critical factors". So, perhaps a good exercise to challenge the misguided confidence in safety that audits can provide, would be to bring together a cohort of operational workers, and ask them what they think leads to the normalisation of unsafe work. Then, take those insights and have a similar conversation with management. I bet you will uncover lots of opportunities and weak signals that years of safety auditing have failed to identify.

I like to think of safety as a social construct in a similar way that workplace cultures are a social construct. It is the people within organisations who create, and continually grow the cultures across their workplaces. And despite most organisations claiming they have one single organisational culture within which all employees are integrated, the fact is that organisations comprise multiple workplace cultures. How could this possibly not be the case? Individual work teams develop cultures that encourage certain behaviours and discourage other behaviours – as do organisational cultures. And within those cultures, safety holds a value alongside all the other competing objectives of the team and the organisation. Sometimes, safety is valued more within organisation-wide cultures than it is in certain team cultures. And sometimes, it is the other way round. The important thing for safety professionals to acknowledge though is that employees are socialised through both organisational culture and work team cultures, so it is super important to gain an understanding of the multiple cultures that exist within your organisation, for the purposes of discovering the value placed on safety at the team level. This is where the real work occurs, and this is where safety professionals should be spending a significant amount of their time. Get to know your people and seek to understand how different work teams contextualise safety and for what reasons. Your goal shouldn't be to immediately change how they view safety – you want to start by understanding how safety is incorporated into their work and why it is more important to some teams and less important to others. Try to gain an understanding of how employees are socialised into these work teams, because the socialisation of employees into work teams or groups, particularly new employees, impacts the cohesiveness and homogeneity of the group, therefore having a direct impact on overall group performance (Gardner et al., 2022). And sometimes, the outcomes of organisational socialisation and work team socialisation can be conflicting, making the experience for new employees (and existing employees) all the more difficult to navigate. When faced with such a conflict, which set of attitudes and behaviours do you think the new employees will assimilate into?

So, just as workplace cultures are socially constructed, safety climates are also socially constructed. I don't see how anyone could argue that safety, to some extent at least, is not a social construct. The meaning people place on safety, and the knowledge gleaned from it, are absolutely socially constructed.

Recognising that safety meaning and safety knowledge is socially created within organisations can be a powerful realisation for any leadership team, often resulting in an 'aha moment' where leaders realise the power of listening to the workforce and building safety programs around employees' lived experiences and shared assumptions of "what is safe". Because sometimes we can all get caught up in the vast library of decontextualised safety models that exist out there, failing to learn from the rich conversations that stem from acknowledging safety as a nebulous social construct for which a universal model does not exist.

Remember, organisations are made up of unique people with unique attitudes and personalities, working together on unique challenges that require unique thinking, in environments where social processes and social factors are highly influential. Following those rigid safety models might actually be holding you back. Perhaps it is time to put them aside in place of adopting a four-step approach to humanising safety. Are you with me?

Chapter Summary

- To view safety as a social construct is to recognise that people develop their knowledge of safety in a social context. What we perceive to be 'safe' depends on shared assumptions and experiences rather than 'safety' being an objective reality.
- Understanding that behaviour is influenced by emotions, social processes, sociocultural factors, social comparison, and the presence and behaviours of others, helps us understand each other and helps shift our perspective on workplace safety to a more humanistic space.
- One of the key assumptions of social psychology is that our inner processes serve interpersonal functions.
- Acknowledging the social determinants of workplace safety, and talking openly about them with management, can help bring safer workplaces to the foreground by facilitating richer, more honest conversations.
- Employees are socialised through both organisational culture and work team cultures.
- Recognising that safety meaning and safety knowledge are socially created within organisations can be a powerful realisation for leadership teams.

MICRO-PROJECT #2

Identify and Build Your Safety Coalition

So much of what we do as safety professionals depends on our ability to influence. But with so much positive change to offer, and so few resources, how can we maximise our influence across the organisation?

Chances are your safety team is pretty slim and pretty busy – not nearly big enough to embark on an internal safety campaign, showcasing the why, what, and how behind the next game-changing safety initiative you have up your sleeve, right? Well, that's no problem. Let's just leverage the key influencers that already exist in your organisation. You know the ones – those people that everyone else just seems to really listen to. The ones who have the ear of their co-workers or whom others just tend to follow for what could be a raft of different reasons. These people aren't necessarily in management positions. In fact, some of the best untapped resources likely exist within operational teams, where the real work happens. These are the people who have the power to make or break your next safety initiative, so let's figure out who they are, let's give them some skin in the game, and let's bring them into your circle! It's time to build your safety coalition.

STEP 1

Take a few moments to sit down and deliberately think about how your organisation is put together. Having an organisation chart in front of you might help. Look at the business groups, divisions, and departments. Look at the reporting lines, the departments that get on well with other departments and the teams that seem to have a bit of a rub with other teams, and then start thinking about the power dynamics that play out across your organisation. Think about the obvious power dynamics and think about the less obvious power dynamics. Formalised hierarchies and reporting lines will present apparent distributions of authority but try to spend a bit of time looking within divisions and teams for those subtle power dynamics. Those nuanced pockets where the influence, privilege, and communication styles of certain individuals seem to show up. Who are the natural leaders in the various teams across your organisation and how influential to the businesses' direction and strategy are these teams? For example, you might have a capital delivery or construction arm in your organisation, an operations arm, a corporate services business group, a human resources department and a media or internal communications division. There will be influencers in each of these areas that you will want to identify and whose energy you want to harness, however from a safety standpoint, the construction and operations business groups are going to be the areas safety plays in the most. Look into these areas and carefully hand pick those individuals who seem to have the ear of their co-workers. The people who others seem to listen to and take guidance from. These people need to become your safety influencers. They will be key players in the success of your safety program, regardless of whether you identify them and bring them into your circle or not. It will be these influential people who others look to when asked to buy-in to the next safety initiative or fob it off. These people can either be your change champions or your resistance generals. It is up to you.

STEP 2

Once you have identified your safety influencers from across the organisation, it's time to activate them. These people will become your early adopters and will turn out to be incredibly valuable resources. This can be tricky and could take all your social influencing skills talked about in Chapter 7. Remember, safety is a social construct, so you need to activate those key players who influence the direction of this construct within your organisation. It's possible you might not have had much to do with some of these people in the past, perhaps you don't seem to gel with them, or you may have even had run-ins with some of them! Try to humble yourself and look past this, keeping in mind these people could be the difference between the success or failure of your safety program. I bet you will find that these people actually become some of your favourite colleagues once you put in the effort to get to know them!

The first step might just be reaching out to them and showing a genuine interest in what they do. The better you can understand how they approach their work, from both a professional perspective and a social perspective, the better you will be able to utilise their standing in their team and in the organisation. If you have a safety team around you, you might want to adopt a *divide-and-conquer* approach. Undertake this exercise together and put in the effort to build relationships with these influencers. You will need to spend a significant amount of time in the field understanding how they work and learning about their team dynamics. At this stage, you are not requesting anything of them, you are simply building rapport and seeking to understand what they do, how they do it, what they need from the organisation, and what's important to them.

STEP 3

Invest in this group of influential, soon-to-be early adopters of your safety program. Give them access to training programs around elements of your eclectic approach to safety and help raise their profile in the organisation. As a safety professional, it is likely you have good working relationships with members of senior management. Lean on these relationships to introduce your influencers to the management group and once you feel your influencers are engaged and activated (noting this could take weeks or it could take months), you can start bouncing concepts off them and gauging their views on some of the big ideas you have for improving safety at your organisation. At this point, you want to be showing them how much you value their perspectives and feedback, asking if they can think of better ways to approach safety issues or challenges and showing them that their input directly shapes the direction of the safety initiatives you have planned. The gold nuggets in this exercise will start to emerge for you at this point. Not only will you be better placed as a safety professional by being closer to the work and building better relationships with the frontline, but you will also find that these influencers will start suggesting you speak with others in the organisation to better understand the frustrations or issues they are highlighting, or who might have deeper safety insights for you. They might even point you to a couple of other key influencers you hadn't considered!

There are so many things you can learn from frontline workers that will never be realised sitting behind a desk and it is through these people that the quality of your learning will be determined. These influencers are your gateway to the workforce. They will become your guiding coalition and champions of change when rolling out new or improved safety programs. Organisational resistance will be overcome by this coalition as they utilise their collective influence to pave the pathway to broader acceptance. Their value to the success of your safety programs cannot be understated – it is your job to show them this. Your organisation's safety vision needs to resonate with them because they represent the needs of your customers – the workforce.

In one organisation I worked for over a decade ago, I adopted a bit of a unique approach to building my safety coalition. I was the safety manager on the client side of a pretty significant construction project, and I had been provided with only a small team of five safety advisors. Knowing we had upwards of ten principal contractors delivering multidisciplinary land and marine-based construction projects, five safety advisors just wasn't going to be enough to roll out our human-centric safety program the way I had wanted to. So, I approached senior management with a proposal. After pitching my ideas for the project-wide safety program I wanted to deliver, and knowing that I wasn't going to get any more safety resources, I asked if management would consider a safety secondment program of sorts, where a handful of passionate and influential employees in other areas of the project team could be offered a secondment into my safety team for just one day a week over a three month period. Management was intrigued and supported the program, so we ventured into the project team looking for five passionate employees from five different teams, all of whom had expressed an interest in development opportunities within the organisation. The safety secondment program was framed by the internal communications team as a development opportunity that senior management had designed and were committed to, for the purposes of providing development opportunities to those who were interested in progressing through the ranks of the organisation (it wasn't stated quite that way but this was the underlying message). I was younger then, and at first, I was a little discouraged that my great idea had been snapped up and re-badged as management's brain child, however once I got over my ego, things could not have gone better! We had an enormous response to the expression of interest, and it was actually quite difficult to select just five people. But once we had our extended safety team, we welcomed them, put them through some internal training, upskilled them in human-centric safety principles, coached them on how to have safety conversations with our contractors, and then after three months, we gave them back to their original teams. The thing is, we never really completely let go of these people. They remained engaged in our safety program and became a shadow extension of our small safety team, embedded in different departments across the project. The program was so successful that we repeated these three-month part-time secondments for a further three rounds. In the end, my modest team of five safety advisors grew to 20 passionate, safety influencers!

Now that you have identified your safety coalition, keep working on those relationships and keep them engaged on safety matters. Like all relationships, everything is built upon a foundation of trust and reciprocity, so you will also need to understand what they need to help make their work easier and more productive, utilising your influence with management to make things happen for their benefit. The quality of your relationships with them, together with their collective influence throughout your organisation, will prove to be critical in applying the four-step approach to humanising safety.

References

Baumeister, R.F. (2020). *Social Psychology and Human Nature.* Cengage Learning.

Gardner, D.G., Huang, G. (Emily), Pierce, J.L., Niu, X. (Peter), and Lee, C. (2022). Not just for Newcomers: Organizational Socialization, Employee Adjustment and Experience, and Growth in Organization-Based Self-Esteem. *Human Resource Development Quarterly, 33*(3), 297–319. doi: 10.1002/hrdq.21458

Hutchinson, B., Dekker, S., and Rae, A. (2024). How Audits Fail According to Accident Investigations: A Counterfactual Logic Analysis. *Process Safety Progress,* 2024, 1–14. doi: 10.1002/prs.12579

Lynch, C. (2018). *Identity Crisis: Reclaiming a Cohesive Identity on a Nursing Education Unit*. ProQuest Dissertations Publishing.

Turner, N., and Gray, G. (2009). Socially Constructing Safety. *Human Relations, 62,* 1259–1266. doi: 10.1177/0018726709339863

Part 3

A Four Step Approach to Humanising Safety

We've been through quite a bit in our short time together, you and I. We have covered a considerable amount of ground from talking about the challenges associated with defining safety and some differences between traditional and progressive safety approaches, through to the human condition, the five perspectives of psychology, our relationship with failure, and safety as a social construct. And although there is so much more to explore and unpack around the psychology of safety, I feel like enough groundwork has been laid to allow us to move into my four-step approach to humanising safety. And you know what? I'm not even going to drag this out! I won't be employing narrative tension or other cunning methods to keep you peaking around the corner of the next page to learn what these four steps are (just in case you didn't pick them up from the contents page!). I'm just going to give them to you straight up, and then I will spend some time exploring these concepts, using examples and case studies that have been shared with me by a handful of outstanding, generous safety leaders. Just a heads up though – these four steps are by no means earth-shattering or radical. They are actually very simple. Some might even argue they are just common sense. Possibly not even worthy of wrapping a book around them! No – kidding. Completely worthy of a book! So, without further ado, here are the four steps to humanising safety:

1. Get to know your people
2. Make safety simple
3. Involve end users in the design of safety programs
4. Crowd source for safety solutions.

DOI: 10.1201/9781032679525-10

Pretty simple right? But before you put this book down to charge out there to put these steps into action, I would encourage you to stay with me and read on, because there are plenty of practical tips across the following chapters that will help you maximise your chances of success in applying a human-centric approach to safety at your organisation. And there are also some pretty awesome wisdom bombs embedded in future chapters from thought safety leaders and heavy hitters such as Josh Bryant, Kim Bancroft, Brad Green and Zoe Nation. I hope you get as much out of these four steps as I have.

8

Step 1 – Get to Know Your People

A desk is a dangerous place from which to view the world.

John le Carré

The people in your organisation are your biggest asset. I bet you have heard sayings like this loads of times, right? Organisations employ skilled people to perform skilled work and clever thinkers to help solve problems, build strategies, and stay ahead of the competition. But so often these skilled, clever workers are not provided the right conditions to help them truly thrive within the organisations they work. So often they are restrained by organisational drag, convoluted processes, internal politics, or bureaucracy, making it difficult for them to successfully deliver what it is they were employed to deliver. Their skills and knowledge are not utilised by their employers. As this untapped library of knowledge builds among the workforce over time, divisions or silos between management and the workforce are formed and strengthened, and language like "us and them" begins to emerge. Sub-cultures are formed and espoused company values are met with cynicism and resentment. Management doesn't know the workforce and the workforce doesn't know management.

When these types of environments are inadvertently created within an organisation they can be very difficult to dismantle, and quite destructive for safety teams to navigate. Any attempts at rolling-out new safety programs are stifled because workers feel like safety is being done *to* them rather than *with* them. "The safety team doesn't know who we are. They're part of management. They don't understand the challenges we face every day. How could they? We rarely see them!"

As stated in the preface of this book, I have the benefit of having worked 'on the tools' for a bit over a decade in some really diverse and high-risk industries. Civil construction, mining, offshore oil and gas, maritime, aviation, demolition, and utilities. I have been in 'camp worker' and for the later part of my career I have been in 'camp management.' I have witnessed (and played into) the division that forms between both groups, but thanks to my experiences, I have also discovered a very effective way – a very simple way – to start to break some of these walls down. All you need to do is get to know your people. I know what you are thinking. It sounds too easy, right? And it is!

DOI: 10.1201/9781032679525-11

I am not going to labour on this step too much because I don't want to tell you how to suck an egg (side note: Google where "suck eggs" came from. It really is quite strange). I'm sure you have plenty of great working relationships with the employees of your organisation and I understand you know perfectly well how to build relationships with people. But I wonder how much you really know about your workforce. Are your relationships mostly surface relationships that do not extend much beyond saying "hello" as you walk by and perhaps asking about the weekend or giving someone a bit of a ribbing over Friday nights' footy game? Or are they deeper relationships, like friendships, built on trust and reciprocity, where you really know what is important to them, and where you deeply understand what they are passionate about? Do you know what energises them and what de-energises them? Do you know what they are looking forward to or saving up money for? Do you make the time and effort to regularly listen to some of the shit they are going through both at work and in their personal lives? I know if you stand back and look at all the employees in your organisation, it might seem like an impossible task to get to know everyone on this level. The key is to just make a start. Crawl, walk, run. Get away from your desk as much as you can and try to embed yourself in the various operational teams that exist out there in the field (or in the office). Venture outside, show a genuine interest in others, practice active listening, and before you know it things will start really coming together for you. Because the success of your safety programs relies on the quality of the relationships you hold with those closest to the work. They are the ones who will make it or break it for you, and if you have made the effort to form good quality relationships with a decent cohort of the organisation, then you are going to get much less resistance when it comes to that next safety initiative.

I remember joining one organisation as the Head of Safety, tasked with leading quite a large team of 22 enthusiastic safety advisors to pave the way for a new direction loosely built on the HOP philosophy. The challenge facing me was to rebuild the organisations' safety approach from behavioural-based safety to humanistic safety, after many years operating in the former. The safety team was well established in the organisation and well respected by management, and it was obvious that these safety advisors had spent a decent amount of time getting to know middle and upper management. However, it became evident in my first few weeks that the same level of effort in building relationships had not been exercised with the operational field teams. And if there were to be any chance of successfully facilitating a step change in safety, then this team of safety advisors was going to have to get busy getting to know the people who made up the majority of the organisation!

So, after gaining approval from senior management, we paused our safety program for three whole months, just to get to know our people. The only safety tasks we performed were those critical BAU elements. I assigned each safety team member (including myself) one operational team each, and we essentially became a member of that team for the month. We turned up to work at the same time as them, we participated in pre-start meetings and toolboxes, we received basic task-based training, and for a period of one month, we undertook the same tasks that others in that operational team performed. We became an additional member of the team and through this exercise, we really got to know each other, and we received invaluable insights into the sub-cultures that existed in the business, the difficult and dangerous elements of their work, and the over-burdened work processes and bureaucracy that constrained them. And probably most importantly, we experienced those impractical and downright dementing elements in the organisations' safety management system that the safety team had been demanding of them for years!

After the month was up, we all rotated to a different operational team and we did this over a three month period. It was an incredibly humbling and valuable exercise and one that given the chance, I would highly recommend you attempt in some way, shape or form (appreciating that many organisations may not have the luxury of freeing up their safety teams for long periods of time). In fact, as I am writing this chapter, the safety manager in the organisation I currently work with, and who reports to me, is about to embark on a three-month secondment into an operations manager role. I'm convinced the insights he will gain and the relationships he will form or strengthen from this secondment will directly compliment and benefit his standing as a safety manager when he returns to the role. I know this because when I worked in the aviation industry as the Head of Safety, I sought a six-month secondment into an airfield operations manager function to help me broaden my understanding of the operational challenges faced by employees and to fast track my pathway to getting to know these workers.

Less demanding variations of these frontline assimilation exercises could include spending one day per week/fortnight working as a frontline employee or selecting a couple of operational activities to be involved in as a peer from time to time.

No matter how time-poor any safety professional is, spending time with the frontline, getting to know your workers and the everyday challenges they face, is an element of the safety professional's role that must be prioritised. Because in my experience, the most exceptional safety leaders are those who make the time and effort to really get to know the people doing the work. Safety is built on relationships, relationships require trust, and trust is built through connection. So, get out there and start connecting!

Example 1: How Josh Bryant Does It

Josh Bryant is a senior executive with a passion for learning, people, safety, and risk management. With over 20 years of diverse experience in safety and operational learning across several industries, Josh is known for leading with care and relies on his knack for building strong relationships with people to help him deliver organisational safety programs. Being a well-recognised and widely respected safety leader in Australia, I asked Josh if he could share with me the importance he places on forming relationships with the folks on the frontline, and how he shows genuine curiosity and empathy, together with an example of how he goes about it. He kindly obliged. This is Josh's approach:

"Why would I give them positive feedback for doing their job? That's what they are paid to do – their job! I'll give them feedback when they stuff up." said 'Coxy', a key supervisor in the middle of our internal leadership program. As I put my head in my hands, I stopped and looked up and just said to him "Do you have a dog? Do you give it a pat, just because it's your dog? And you care about it?" "Of course I do!" replied Coxy. And in that reflection, I spoke to Coxy and the other supervisors in the room about how the safety team 'shows up' – and it's something (as a reader of this book) that we all need to reflect upon. Is it only when something 'wrong' happens or showing up to an 'audit' or a check? Or do we show up to learn and be interested? That discussion with Coxy led to a lot of what we do as a team today – self-reflection can take a lot of personal courage. We need to be there in the absence of failure, working to help the business understand how our people make us successful.

If you are a safety professional, or a leader, or both, how do you show up to the job site? Do you show up to catch them out, or do you show up to be genuinely curious and to learn? We honestly used to show up to ensure compliance and have now shifted towards learning. I've had to overcome my own initial biases, and through this I've found that our frontline workers have so much practical and detailed technical knowledge that would outweigh a university degree.

Where to start? Not all of us are good at it, but I'll usually start with small talk. But not something general – more along the lines of their commute to work, how do they deal with conflicts of family and work, what's their experience with interactions with new people to site, have things changed since they joined the industry, how do they switch off from work at home or at the remote site? Rebecca how often do you call your wife? Is it every night, how do you cope with the fly in fly out? - and then genuinely listening for the response to find a connection between us. And once I have that, I will then ask about their work – if it's appropriate.

Rob Fisher, a mentor of mine in the US, uses the technique of "TEDS' – *Tell me, Explain it to me, Describe it to me, Show me.*

Rob also quoted to me that 'management and safety do an awful lot of time-in-field programs, but I ask them 'what are the top 5 improvements to work that programs have done?' Many can't answer that (even most of our clients can't) – and it's because we seek to instantly change other's behaviours first, whereas the curious questions we need to be asking should be about the context they work in and conditions they face.

I mainly work in mining, and the clients' graduate engineers participate in a job rotation on shift so they can understand the work – this should apply to all industries. See if you can do the same in your business – don't just observe - try and do. They will appreciate you discussing all the administrative work it took to even be allowed to do it, and the training sign off, all the hurdles you had to go through – that opens a door to discussion and connection. We've had senior leaders (general managers) work on a site for two weeks as part of the work crew doing what an entry level person would do, and to go through the frustrations of being a new starter. I've had operational managers call me after I've facilitated a Learning Team with the comment *'we know through your actions and response that your skin in the game is to ensure that we have a voice.'* That's pretty cool feedback to a safety leader.

A poignant moment for me personally was when there was a workplace accident where Sebastian, who was only 25 at the time, had his hands crushed in a piece of machinery. My initial response (inside voice) was *'how was this able to even happen?'* However, when first speaking to him I didn't do the usual *'are you ok?'*, instead I genuinely asked about his partner, any upcoming plans, any impacted hobbies or commitments, anything that he needed, before even going near the event. And because the work team were used to the safety team being onsite and being curious about improving the 'safety of work' all of the time, he said that it always felt that *'I had his back – not against the client or as an employee, just genuinely had his best interests in mind'.* That was only possible by being genuinely curious and showing empathy.

Our safety program now only focuses on critical risk management, how leaders respond to failure (Human and Organisational Performance), and organisational learning. We wouldn't have been able to implement critical risk management without the relationship with our frontline workforce and involving them in its design. Nor Human and Organisational Performance if we didn't have the relationship with our supervisors – particularly with supervisors like 'Coxy' who feel that it is ok to challenge (and who now gives his crew a lot more positive feedback).

'Safety' (building capacity) applies to all industries - we have by no means 'solved everything'. We've improved our controls and conditions using tools like *Learning Teams Inc's 4D's* (tell me what's dumb, difficult, dangerous or different) and that has only been possible through non-transactional conversations. You can't fix the work if you don't understand it, and you're not going to understand it without building a rapport with those who do the work – and through their open conversations we've improved the safety of work leading to reductions in our injury severity and halving the number of injuries. But these numbers are nothing compared to the reduction in the number of people giving feedback in exit interviews that they leave the business because of poor leadership – because leadership shows genuine interest and concern.

As leaders we will still sometimes shift to blame as a first reaction, but through building relationships and understanding with our workforce, our response has been to ask *'why did it make sense for them to do it like that?'* – and then we go and find out.

And my final tip - never call anyone by their nickname, until you have permission to. I know everyone else may call them *'Kookaburra'* or *'Grizz'*, but let them be the one to tell you that it's ok to use it. Always start with their first name and continue that respect until you gain trust.

If you would like to learn more about Josh's approach and the 4 D's he mentioned, then do yourself a favour by grabbing a copy of the book he co-authored with Brent Sutton, Jefferey Lyth, and Brent Robinson: *4Ds for HOP and Learning Teams: A practical how-to guide to facilitate learning from everyday work, critical and dynamic risks with the 4Ds.*

Chapter Summary

- The people in your organisation are your biggest asset. Help provide the conditions to allow them to thrive.
- You can't fix the work if you don't understand it, and you're not going to understand it without building a rapport with those who undertake it.
- Getting to really know your people is the first step to breaking down divisions between management and the workforce.
- Are your working relationships mainly surface relationships or are they deeper relationships based on trust and reciprocity?
- The success of your safety programs relies on the quality of the relationships you hold with those closest to the work.
- "Us and them" sub-cultures between management and the workforce stifle safety programs due to workers feel like safety is being done *to* them rather than *with* them.
- Embedding yourself in frontline teams for a period of time to contribute as a peer, is an effective way to build relationships with the workers and understand how they navigate the challenges of everyday work.
- Blame can often be a first reaction, however, you can change your response by building relationships and understanding with your workforce, asking yourself *'why did it make sense for them to do it like that?'*

MICRO-PROJECT #3

Learn about the People in Your Organisation

This micro-project is all about getting away from your desk and making the time and effort to build deeper relationships with the people of your organisation. There are loads of ways to do this and depending on the context of your organisation and the work it does, you may have to vary this micro-project to make sure it aligns with how your organisation operates. Remember, these micro-projects are just idea-generators, so if they prompt you to think of different ways to achieve similar outcomes, then I would absolutely encourage you to design your own approach. So, with that said, here are some short idea-generators for you!

STEP 1

Similar to Micro-Project #2, pull out your organisational chart (if you need to) and have a good look at how the various groups, divisions, departments and teams are put together. Identify the teams that are closest to the work – field crews, maintenance teams, operational departments, construction crews, shop floor teams, supervisors and team leaders. After identifying these groups, ask yourself how much time you have spent with the people in these teams and how well you know them. How well you really know them. A simple way to do this is to ask yourself about their hobbies or interests outside of work and whether you know what is important to them. If you cannot answer this question, then you're operating on surface relationships only and you need to roll your sleeves up and really get to know these people.

STEP 2

Once you've identified the teams you need to invest more time and energy into (and this may be several teams!), start thinking about how you could spend time with them in a way that doesn't impede their work or that isn't completely awkward for you or them. For example, you don't want to suddenly start poking around in their work picking up hazards, doing safety inspections, grilling them on past safety incidents, or heaven forbid, performing behavioural-based safety observations! This isn't how you get to know people. You will get to know how they behave when they're pissed-off though! Instead, take safety out of it and just focus on trying to understand who they are as human beings. As Josh said, you want to ask questions to show genuine curiosity and empathy. You want to learn about them – what excites or motivates them, what they value in their personal lives, their loved ones, who they rely on, what annoys them, and how you can show up for them. It really is no different from how you form friendships outside of work (or inside of work). You can lean on your safety coalition to help facilitate this exercise if needed. That team of safety influencers can be your access to the workforce. Ask them if you can accompany them to a job site or visit them whilst undertaking an operational task, to learn about their co-workers and to learn about normal, everyday work.

Start small. See if you can assign a couple of hours a week in the field just getting to know these people, and then build that up to half a day or a full day as part of your weekly routine. Trust me – you won't regret it! And if your organisation allows for a more comprehensive in-field rotation program like the one explained earlier in this chapter, then jump at the opportunity as it will gift you more learnings about your people and the challenges they face than you could otherwise glean from years of desk-bound work!

STEP 3

If you are anything like me, you might be a little forgetful from time to time. That's ok – you are human after all and your heuristics or cognitive shortcuts, like mine, are probably just highly active, saving your brain's limited energy. The trouble I find with forgetfulness though, is that I tend to lose people's names almost immediately after meeting them. It is something I have battled my whole life. Everyone becomes 'mate' unless I make a conscious effort to commit their names to memory. You can see how this might be a problem, right? This is a very important aspect of getting to know your people – you need to remember their names and you need to actively listen to what they tell you and show them that you are actively listening. A simple method I use is to jot their name, position and a few key points down in my pocket notepad after I have finished an initial conversation with them. Nothing exhaustive. Just enough to help me keep track of my conversations in those early stages of getting to know people until those basic details are committed to memory. It might be nothing more than: *'Sam. Excavator operator. Golf tragic, just bought new clubs. Frustrated by not having dedicated digger (would take pride in own machine)'*. With a simple note like this about Sam, I have all I need to continue the conversation the next time I see her and to show that I was listening (and interested) in what she had to tell me last time. For me, keeping notes helps a lot when I am out and about in the field forming relationships. Especially when I have a long list of colleagues to get to know!

9

Step 2 – Make Safety Simple

How can you govern a country which has 246 varieties of cheese?

Charles De Gaulle

In a great paper by Rae et al. (2018), the authors defined 'safety clutter' as *"the accumulation of safety procedures, documents, roles and activities that are performed in the name of safety, but do not contribute to the safety of operations."* I think everyone who has been active in the workforce at some stage or another over the past 20 years can relate to this. Safety is 'overcooked' in most organisations. The workforce is drowning in safety clutter. Sadly, I am embarrassed to tell people that I am a safety professional when they ask me in general conversation what I do. I am not proud of what the safety profession has become. When I do tell people that I work in the safety field, I find myself following this up with a 10-minute explanation of how my approach is different from the compliance-focussed, 'police officer' mentality most people typically associate with workplace safety. WHS does not have a good reputation in Australia. And for good reason! Rigid safety rules, duplication, over-prescriptive procedures, bloated safety management systems, surface-level audit findings, pointless tick-and-flick exercises, mind-numbing training and induction programs – do we really need all this stuff? Does it really add value to an organisation or increase organisational safety? Is the workforce buying what management is selling? No, of course not. Most of this stuff is just window dressing. A fear-driven response to safety regulations and a distraction from the real issues threatening good operational management (of which safety is just one element). Safety clutter is an unproductive misuse of time not just for the safety teams designing, facilitating, writing and rolling this stuff out, but also for the frontline workers, supervisors and managers who are unduly expected to follow every safety rule, and apply every overly prescriptive safety requirement to their daily work. This just isn't realistic. Safety is far too complicated than it needs to be. Ask yourself, what do people do when something is too complicated? They find another way that makes sense to them, right? They look for shortcuts (if they can get away with them). If a work process is excessively convoluted, people don't believe the reasoning for the complexity is justified, and there are other options available to reach the same outcome – simpler options – then they will take the path of least resistance. It's human nature. We are hard-wired

DOI: 10.1201/9781032679525-12

to take shortcuts – to conserve our energy. As a safety profession we should stop fighting this inclination by stuffing any empty page with word vomit and expecting our workers to blindly comply with unreasonable, impractical safety requirements. Instead, we would benefit from embracing the fact that the workforce is time-poor and our workers regularly suffer from goal conflicts, made worse by the relentless safety clutter our profession continues to force-feed them. This is not the pathway to operational safety. This is the fast track to employee disengagement, cynicism and distrust.

So, what is behind this urge to produce safety clutter? Is it driven by safety regulators? Is safety clutter a response to how organisations or safety teams have interpreted safety legislation? In part, I would say yes. This has definitely contributed to the countless papercuts imposed by safety teams across the world and the resultant haemorrhaging among the safety profession, making it even more difficult for people like me to admit to strangers that I work in safety. There seems to be an unfortunate common belief among organisations that if they have piles of documented procedures, checklists, rules and work instructions that cover every imaginable scenario, then they increase their legislative compliance and in doing so, they have built a solid legal defence in the event of possible prosecution. But safety clutter does not translate to legal compliance. Safety clutter is not due diligence. Managing work safely is.

Safety legislation in Australia requires employers to eliminate or reduce safety risks and provide safe work environments for their employees. This extends to safe equipment, safe systems of work and ensuring workers are trained and competent to undertake the tasks requested of them. Sure, there is a bit more to it than just this, but if I was asked to summarise it in a sentence or two, I reckon this is how I would respond. In the event of a serious workplace injury or fatality, do you think an organisations' safety clutter will defend them in court? Do you even think it is possible to defend a charge of failing to provide a safe workplace if someone was seriously injured or killed? If someone was seriously injured, then the workplace wasn't safe, right? If it were safe, then they wouldn't have been injured. So, why do we spend so much time and effort on safety clutter knowing it doesn't benefit the workforce and it won't defend our organisations when someone gets seriously hurt? Why is safety legislation interpreted in a way that results in administrative overkill and treating employees like brainless, emotionless, robots? We need to change our approach. Actions are more effective than words and actions are preceded by attitudes.

Another contributor to safety clutter is the additive nature of safety management systems. Sure, the hierarchy of controls asks for 'above the line' controls rather than administrative controls, but 9 times out of 10, audits or incident investigations will be followed by a new checklist, more safety training or something additional stuffed into a safety procedure. This is the low-lying fruit that allows organisations to satisfy themselves that they have appropriately responded to audits or adequately investigated incidents. But for the end user – the people closest to the work – this just adds to the pile of safety bullshit they are expected to wade through and amplifies goal conflicts between safety and production.

Safety clutter tends to build more quickly in organisations whose safety teams have not taken the time and effort to get to know their people. If they knew their people, understood the challenges faced by frontline workers and supervisors, and respected how they accommodate safety into their work, then they would refrain from undermining that autonomy and safety ownership by loading them up on safety clutter. A common safety mantra that rubs me up the wrong way and lives and feeds on safety clutter is the one that states each person is responsible for safety. You will be familiar with these – *"You are responsible for safety on this site"*, or *"Safety starts with you."* What a load of rubbish! How can someone who has little to no say in how safety is managed at an organisational level, be expected to take ownership of it? Particularly in organisations where management (including the safety team) do not know their people. This is a classic case of safety being done *to* people rather than *with* them, and if safety is being done *to* someone, then they are very unlikely to take ownership of it.

In addition to encouraging workers to take shortcuts, increasing goal conflicts and reducing ownership of safety, safety clutter also disengages the workforce, erodes trust in management and safety teams, and makes it very difficult for workers to safely adapt their methodologies to the varying nature of everyday work. The rigidity that comes with safety clutter makes this almost impossible. There is no oxygen around safety anymore. It is suffocating. The truth is that safety clutter is a risk to organisational safety in and of itself, and a misrepresentation of the safety of work within organisations. And tackling it is the second step in my four-step approach to humanising safety.

So, how does an organisation go about making safety simple? Knowing the rapacious nature of safety clutter and how quickly it accumulates in organisations, this seems like an impossible task to so many safety professionals. But I don't think it is too difficult to start to simplify safety. Recall the 'crawl, walk, run' analogy. Just make a start. Starting is enough. Here are a few ways to declutter that have worked for me in the past:

- Start using the term 'safety clutter' around the office and in meetings. Acknowledge it for what it is and draw management's attention to it as another challenge to overcome on the path to improved operational safety.
- As you continue getting to know your people, ask them about those safety processes or procedures that don't add value. Ask them to tell you what they think are the dumbest or most overcomplicated safety procedures, processes, rules, or requirements. Try to understand things from their perspective. Listen to learn, not to respond. Ask them why these things seem dumb or complicated to them. You can use work insights to help with this if you like.

- Through these conversations, when safety clutter is identified to you, do something about it! See what happens when you get rid of a duplicate procedure or low value safety process. See if anyone notices. Of course, you will need to consider any potential impacts of killing a seemingly pointless procedure before getting all trigger happy, so do some investigation and consultation beforehand. Trust me though – it can be very liberating having a safety system garage sale!
- Change your response to safety audits, actions arising from meetings, and incident investigations. Resist that easy option to introduce a new procedure, add to an existing one or develop more training. Think about those 'above the line' controls to prompt you to come up with engineering actions or work design changes instead of more paperwork. Get creative!

I worked with an organisation once who were bold enough to bin their accredited safety management system (SMS) because they accepted that it was providing no value to their workforce and the only purpose it was serving was to pass accreditation audits. Before we threw this 60,000-word SMS in the bin though, we asked the IT department to run a check on how many people within this organisation of around 1,000 staff, had accessed a document within the SMS over the past three years. IT came back to us with some pretty underwhelming numbers. On average, only 9 to 10 people were accessing the SMS each year so we leaped into an organisation-wide consultation process, seeking to understand what would happen if we got rid of the SMS and replaced it with something that was co-designed with the workforce, made sense to them, and actually added value. It was a rewarding exercise and one that was very successful, increasing employee engagement and trust in management tenfold!

Example 2: How Kym Bancroft Does It

Kym Bancroft is quite the accomplished safety leader. Not only is she the Managing Director of New View Safety, but Kym is also an organisational psychologist, an Executive HSE leader with over 20 years of experience, a keynote speaker and an active contributor to the global HSE profession and industry bodies including a former member of Safe Work Australia. I asked Kym to share her pearls of wisdom on safety clutter and ways she has gone about tackling it within organisations she has worked with. True to her character, Kym was more than happy to help. Here is Kym's approach:

Safety clutter is defined as the accumulation of safety procedures, activities and roles that are performed in the name of safety, but that do not contribute to operational safety (Rae et al., 2018). It involves the proliferation of redundant safety procedures, documentation, or practices that, instead of significantly contributing to actual safety outcomes, create a convoluted and unnecessarily complex system. It's like having too many tools in your toolbox, making it harder to find and use the ones that really matter.

There were two papers that introduced me to the concept of safety clutter, both of which have really stuck with me. The first was the paper, "Safety clutter: the accumulation and persistence of 'safety' work that does not contribute to operational safety" by Rae, Provan, Webster and Dekker. This paper, and others published by Provan and Rae, contains several powerful insights, so much so, I was able to build my whole decluttering strategic initiative from this paper.

The second was a report by Deloitte, titled "Get out of your own way" which suggested that "Australia has a problem, and the colour of that problem is red". The report was referring to the self-imposed bureaucracy and clutter, over and beyond that which is required of organisations under legislation.

Both of these papers highlighted to me the significance of the issue in organisations globally, and the way in which our WHS profession has been socialised and indoctrinated.

The inclination to gather safety clutter often stems from a misguided belief or assumption that more rules, documentation, and complexity equate to better safety. There's a societal and organisational pressure to showcase diligence, adhere to compliance standards, and ward off potential legal consequences. It's as if 'more is better' and the more safety measures we invest in, the safer we appear, even if some of those measures don't actually enhance safety. It's the night-time lullaby of being 'papersafe' that allures us, even well-meaning safety professionals, to continue to create and impose bureaucracy. And let's face it, this issue isn't unique to H&S… other closely aligned professions face the same challenges. It can then be very challenging to speak up and suggest removing the clutter, as the ethical pressure to continue to build is ever present.

The call to streamline safety is necessary to cut through the noise and prioritise what genuinely contributes to frontline risk. Identifying and reducing safety clutter means making processes more efficient, removing unnecessary steps, and applying critical thinking to any further processes or procedures that we are adding.

However, while the concept of decluttering intuitively promises much, the practice is much more challenging. There is often reluctance among safety professionals and organisations to remove safety clutter due to a variety of factors. Some fear potential legal repercussions, thinking that more rules provide a safety net against prosecution. Others may resist change, being accustomed to the familiar, even if it's overly complicated. Additionally, there may be a lack of understanding about the distinction between essential safety measures and unnecessary clutter, reinforcing the belief that more is always better in ensuring safety.

This reluctance was observed on multiple levels when I was working as the Head of Health & Safety at a large water utility. The organisation was undergoing an ambitious safety culture transformation over a period of three years. Clutter was identified through a survey that was rolled out as a pre-transformation measure. The HSR's and frontline leaders spoke with frustration about the excessive bureaucracy and paperwork that was meant to 'keep them safe' but in reality, had become a surface compliance 'tick and flick' exercise. To verify this, and to determine whether the issue was systemic and should be a strategic initiative, we rolled out the safety clutter scorecard throughout the organisation. The results indicated we had a bloated, inaccessible safety management system, and significant frontline frustration with the daily risk assessment process.

Anecdotal feedback from field workers indicated to us that this was a challenge in our organisation. The feedback suggested there was a significant amount of drift from 'Work-as-Imagined' when it came to this risk assessment practice. Stories ranged from people completing them in bulk at the start of the week to ease the administrative burden upon arrival at each site, to people not using them at all and creating a gap with the system. There was a significant amount of checkboxes over multiple pages that saw the worker filling in the form separate to the workgroup and the actual conversation on the job planning and management of associated risks.

It was clear that from an operational perspective, this was driving a lot of unhelpful administrative time and taking people away from the important aspect of having a collaborative conversation about the hazards and how to control them. The health and safety team was keen to learn more about the frustration and improve the quality of field risk assessments using the principles of decluttering designed by David Provan and Drew Rae. To better understand the challenge, we embarked on a three-month ethnography study to examine how the Health & Safety Management System contributed to operational safety, and the relationship between the workers' actions and the elements of the system.

Through this discovery, we found that the pre-task risk assessment was a consistent source of frustration. Feedback included:

"They add no value. Just used to cover your arse."

"It's a long process for a job we do everyday."

"Get rid of a lot of the content that isn't relevant once onsite."

"They stay in the folder. Get rid of it."

"We just talk through the work anyway."

To further triangulate this feedback, we conducted a Safety Clutter Scorecard Survey, aimed at further understanding all stakeholder views on the value (or non-value) of all safety activities in the organisation, including the pre-task risk assessment. Out of 23 safety activities, the pre-task risk assessment was in the bottom 5 of value add. In developing a solution, we used a Human-Centred Design methodology of discovery, creation, implementation, and review. At a high level, these are the steps we followed:

- Structured discussions were had with Voice of Intent (Executive Sponsor), Voice of Experience (frontline workers who use the process) and the Voice of Design (H&S team).
- Data was captured and themed against recurring themes, contradictions, unique ideas, gaps, and problems/opportunities to move ahead with.

These themes were then analysed using the different "voices" to help craft up opportunities into action that help to meet operational objectives. This included a Project Plan with activities to align with objectives from the H&S Strategy. The Learning Team process also helped us to understand the current pre-task risk process from a 'Work-as-Done' perspective, and to design the solution in collaboration with the workers. This gave us rich stories around what parts of the process were helping and which parts were a hindrance. The concept of decluttering the pre-task risk assessment, based on the discovery feedback, was then taken to the Board and the executive for consideration and endorsement. We anticipated that a common objection would be that it would compromise the legislative obligations of the organisation. To place this objection out, we strongly considered the legal obligations in undertaking pre-task risk assessments by engaging with leading H&S Lawyer Michael Tooma, to provide a review and recommendation on a path forward to ensure all legal obligations were being met.

What we discovered in our consultation with legal was that there was no legal requirement to record the risk assessment outside of confined space and diving. There was a benefit when the work was complex and in having confirmation or verification of the additional controls that were put in place. Specifically:

> There is no express obligation to implement documented risk management processes in the WHS Act, Regulations or Codes of Practice. Whatever obligation exists relates to proactive management of the risks. Documentation of that process is only necessary if it assists in traceability of implementation of risk controls and review of the effectiveness of those controls.
>
> **Clyde & Co.**

Following endorsement from the Board and the executive, we then embarked on the design phase. In doing this, we gathered a team of frontline staff and asked them to take a critical look at the existing document. We created a space where they could be frank about the value the process had in helping them manage safety. The team told us that there was very little in the existing process that they use and that they normally have a team meeting before they commence work to discuss. We gave them a red and green marker and asked them to put a red line through what was not helpful and a green tick against what they thought added value. This gave us the idea about creating freedom within a framework for them to utilise.

Through these discussions with the field workers and frontline managers, the "CHAT" was designed. Interestingly, this name was derived from an informal competition run with the frontline groups participating in the roll-out training. The "Conversation Hazard Awareness Tool," a guided tool for assessing and mitigating risk in the field was developed. It consisted of seven questions to risk assess any job that fell into the criteria of basic and routine. It was designed with risk competence in mind, and to be a proactive, structured, and systematic approach to worksite safety that removed administrative burden and aligned 'Work-as-Imagined' with 'Work-as-Done'. To implement the CHAT, a training program was developed that walked the field workers through the change process, the feedback we had received, the design process, the reiteration, and the final solution and how to implement it.

Partnering with Dr Tristan Casey, we designed and delivered a training program that helped to introduce the CHAT. This program covered areas such as the subjectivity of risk assessments, critical conversation skills, and information about in-scope and out-of-scope tasks. We educated the workers in how to apply the CHAT through scenarios that were specific to that team being trained. It included a verification of a competency component which leaders undertook out in the field. A further leaders training program was rolled out equipping them with the skills to verify competency. Post-intervention feedback indicated that this was the most value-adding safety project the workers had experienced during their time in the organisation. They were very grateful that the health & safety team and leaders had listened to their concerns and involved them deeply in the process all the way from creation through to implementation.

Despite removing the requirement to record the pre-task risk assessment, we found that some workers found it beneficial to record it, stating they liked writing it down. A few were concerned about personal liability if something went wrong, and even though they knew it offered nothing in terms of prevention or risk control, liked the 'security' of the written document. So, a written version of the CHAT was retained for workers preferring this. We also considered how we could use technology to assist in building a stronger culture on the pre-task risk assessments. The CHAT was embedded in the work planning process by being attached to work orders that were issued to workers so it was always visible. All the frontline workers had smartphones, so we encouraged them to take photos of the worksite and use text messages, voice-to-text notes, etc. Some groups had created a Microsoft Teams page that allowed them to share the work that was being done. This also allowed the health & safety team to cascade messages across the other groups. Regular project updates were shared with all stakeholders through our reporting on strategic initiatives. When the project concluded, a summary video of the field workers' perceptions was collated and shared throughout the organisation.

Some of the feedback included:

"This was the first time a safety change has involved us in the solution."

"Finally, a safety change that doesn't make our life harder."

"An easy-to-use tool that encourages collaboration on the job."

"I love it, easy to remember and encourages conversation on the stuff that matters."

"Less paperwork and more focus on each other and the job at hand."

The solution felt highly intuitive. It was driven by the frontline workers. We listened to their frustrations and empowered them to build the solution with us, and we took the components forward to the blunt end for endorsement. In retrospect, change of this nature with a well-ingrained practice such as this is challenging. Whilst the problem was clear and the negative impact on operational safety was evident, the resistance to shift to a solution that doesn't 'feel' safe and puts so much trust in the frontline worker is highly threatening to some leaders.

To be successful in this change, it is imperative to work with these stakeholders and allow the time for them to become used to the idea that people are the solution. Mitigations such as regular checks to see that it is working effectively, and helping senior executives feel more comfortable with the change at hand, is imperative. At the end of the day, this change initiative meant that we were upholding safety as an ethical responsibility to the frontline worker, instead of prioritising the bureaucratic activity of recording the risk assessment.

In summary, pre-task risk assessments were found through consultation and collaboration to be hindering operational safety rather than assisting the frontline workers with risk management. Through a process of human-centered design, we were successful in 'decluttering' the process and aligning 'Work-as-Imagined' with the 'Work-as-Done'. The result was a worker-led design, the CHAT, a conversation-based risk assessment that was extremely well received by the frontline workforce.

If you would like to learn more about how Kym Bancroft and Dr Tristan Casey help organisations turn contemporary safety ideas into operational practices, check out their web page at https://thenewview.com.au/

Chapter Summary

- Safety clutter builds quickly and is slow to deconstruct.
- When safety processes are too complicated, the frontline will just do what makes sense to them. This includes taking shortcuts.
- Safety clutter is not the pathway to operational safety, but rather an express route to employee disengagement, cynicism and distrust.
- Many safety professionals and organisations are reluctant to remove safety clutter due to fear of potential legal repercussions, thinking that more rules provide a safety net against prosecution.
- Safety clutter is not due diligence. Working safely is.
- Safety clutter is like having too many tools in your toolbox, making it harder to find and use the ones that really matter.
- Some techniques to address safety clutter include talking about it, asking the frontline to identify examples of it to you, experimenting by getting rid of duplication procedures or low value safety processes, and changing your response to audits, meeting actions and incident investigations.
- There is no express obligation to implement documented risk management processes in the WHS Act, Regulations or Codes of Practice. Whatever obligations do exist, relate to proactive management of the risks.

MICRO-PROJECT #4

Identify and Address One Piece of Safety Clutter

Appreciating the reluctance for many safety professionals to identify and eliminate safety clutter, this micro-project can be approached as an experiment of sorts. A micro-experiment. Approaching the simplification of safety with caution makes sense because as is the case with anything, one can get carried away and inadvertently introduce unintended risks. The goal of this micro-project is to be measured in your approach to simplifying safety by using your safety coalition to identify one or two pieces of safety clutter that are not providing any value, and then taking steps towards eliminating it from your safety system.

STEP 1

Look to your safety coalition and HSRs to help you identify a safety process that is needlessly complicated, does not make sense to them, or a duplicate of another process that has not been attended to in the past. This could be in the form of a procedure, a checklist or inspection, a safety rule, a training module, or a risk assessment process. Gather your safety coalition or approach them individually and tell them about this micro-experiment you are looking to conduct with them, sharing the concept of safety clutter and committing to identifying and eradicating one low value safety artifact for the purposes of observing what impact, if any, this has on operations.

Ask your safety influencers to identify one safety process, procedure, rule or task each that is a pain point for them, or just seems downright impractical or dumb. You will find that in no time at all you will have compiled a lengthy list of potential items of safety clutter. This step can be fun. Try to enjoy it!

STEP 2

Again, with your safety coalition, review the list you have compiled and select one item to form the basis of your experiment. It could be a simple item like a duplicate procedure, or it could be a more complex item like a permitting process. The basis for selection should be that this process, procedure, rule or task is not providing much value in its current form. Once you have selected your suspect piece of safety clutter, engage in an appropriate level of consultation within the business to gain a wider understanding of who it is intended to apply to, how it is being perceived by its intended audience, and whether or not it is providing value to anyone outside of your immediate safety coalition. You want to make sure you are speaking with the intended end-users before making any decisions around amending or eradicating the process, procedure, rule or task. Consulting with your HSRs at this stage is a good idea and depending on your decision-making delegations in the organisation, you may need to consult with your manager too.

STEP 3

With consultation completed and your item of safety clutter confirmed, it is now time to get to the task of simplifying safety. Speak again with those whom this process affects (the end users) and ask them to share with you how they think it could be improved and made more practical. If it is a simple case of a duplicate procedure, the fix will be straightforward and may involve just deleting it. Job done. But if the safety clutter is due to an overcomplicated procedure or a seemingly pointless safety process, you will want to involve the end-users in redesigning it in a way that makes sense to them, addresses frontline risk, and reflects 'work-as-done.' Try to be a facilitator of this exercise and resist the urge to rewrite the process yourself on behalf of everyone you have engaged with. Remember, the end users are the experts in what they do so it is important that any safety process developed for the purposes of increasing the safety of their work, involves them in the design of it. You may find it useful to pull together a process map based on their lived experience of how the activity or process works in practice, you might choose to facilitate a Learning Team with them, or you might engage your safety coalition to help expose the gap between 'Work-as-Imagined' and 'Work-as-Done'. Whatever your approach, make sure it centres around the end users.

STEP 4

Now it is time to go ahead and eliminate that first piece of safety clutter. You have done plenty of preparation work in consulting with HSRs and end users, and unpacking this item of safety clutter with them to understand why it isn't adding value, so you needn't feel like you are being reckless. Remember, adding clutter to your system is easy – removing it is difficult. So back yourself in and hit that delete button! Like with any continuous improvement initiative, you will want to circle back to your safety coalition, HSRs or end users after a month or two to understand how things are going for them with one less piece of safety clutter. Ask them a few questions around whether it has made their work, or that particular task, a little bit easier, or if something may have been overlooked or not realised during the planning phase of this micro-experiment. It is possible that by removing the process, your end users have subsequently realised why it existed in the first place and this exercise has provided insights that would have otherwise been missed. Use these insights to re-design the process with them in a way that makes sense to them and clearly captures the missing intent these insights have highlighted. If this is the outcome, consider it a win also.

If, like you hypothesised, the removal of this safety clutter has not resulted in any detrimental impacts to the frontline, try to capture this through conversations with HSRs and end users. Management being management, will likely ask for measurements of success or evidence that legal exposure has not been increased by removing this process, procedure or safety activity, so try to extract some of this information from your safety coalition when conducting a post-elimination review. Use quotes from the frontline if they provide them. It can be very powerful to present statements like *"removing this process has allowed me to focus on the actual risks"* or *"we all knew this tick-and-flick exercise was there just to cover management. It's good to see our feedback is finally being heard."* Any leadership team or executive group acting with even a shred of integrity will not dismiss quotes from the frontline such as these. Direct quotes are gold! You should actively seek them out. Remember, one of the critical aspects of the safety professional's role is to give the workforce a voice at the executive table. This exercise will allow you to do just that. Share the findings from this micro-experiment with management, share post-elimination learnings and quotes from the frontline, reiterate to management that the business cogs did not stop turning as a result of this exercise (nothing blew up and no one got hurt), and importantly, use this micro-experiment to gain traction and trust from management so you can replicate this process with more items of safety clutter. If you can manage all of this, then you are well and truly on your way to making safety simple.

Reference

Rae, A.J., Provan, D.J., Weber, D.E., and Dekker, S.W.A. (2018). Safety clutter: the accumulation and persistence of 'safety' work that does not contribute to operational safety. *Policy and Practice in Health and Safety, 16*(2), 194–211. https://doi.org/10.1080/14773996.2018.1491147

10

Step 3 – Involve End Users in the Design of Safety Programs

> **People who think they know everything are a great annoyance to those of us who do.**
>
> **Isaac Asimov**

Have you ever stopped to consider who the customers of your safety programs are? Who are the beneficiaries of your safety initiatives and who would be disadvantaged if safety management was not an element of your organisation's operating model? I find reflective questions like these super helpful and an effective way to re-instate perspective or ground me in times when the bureaucracy of safety starts weighing me down. Whilst the customers of an organisation are very clearly known, understood, and considered in the way the business operates, it can be ambiguous for the safety professional to identify his or her customers, and this can present significant problems when trying to make value-based decisions around the needs of the customer. It is a little like forging forward on a safety management assignment without a clear definition of safety in the context of your organisation (refer to Part 1: What is safety?). How can safety professionals make decisions or design safety programs for their customers if they don't really know who their customers are?

I am a firm believer that the workforce should be viewed as customers to the safety programs an organisation develops and rolls out. Those frontline workers, supervisors, team leaders and managers are the end users of whatever safety programs or initiatives a safety team or management pulls together. They are your customers. This is undeniable. They are the people you are appealing to and whom you want to recognise a need to use your product (your safety programs). Without them, you don't even have a product, so doesn't it make sense to involve them in the design of your products – in the design of your safety programs?

Early in my safety career, I viewed senior management as the customers to an organisations' safety efforts, focussing my time and energy as a safety professional on meeting their needs and doing everything I could to protect their interests through building an organisation that was 'paper safe'. But I came to realise that developing my products to meet the needs of management felt unauthentic and was a disservice to those workers on the frontline,

DOI: 10.1201/9781032679525-13

whose roles became increasingly more difficult with every management decision about safety that was made on their behalf and without their input. So, I changed my approach, bringing management into my 'workshop' as supporters and sponsors of the safety programs we rolled out together, which were designed and built in response to learning directly from our real customers – the workforce.

Involving end users in the design of safety programs is step three in the process of humanising safety and probably what I would consider to be among the key determining factors of success for any safety professional.

If you put yourself in the shoes of the frontline worker – or even better, if you have experience working at the coalface yourself – you will recognise that they tend to inherit trouble from upstream. Leaders and management whilst doing the best they can and with good intent, will occasionally (or frequently) make decisions that fail to recognise how the frontline might be impacted. The competing pressures faced by leaders, such as budgets, schedules, resourcing challenges, shareholder or customer expectations, or commitments inherited from the executive and Board, can be really difficult to navigate. These things shape their decision making and sometimes, poor old workplace safety is inadvertently (or sometimes knowingly) overlooked, ultimately increasing frontline risk and allowing trust in management to erode.

One way to overcome this impact on the frontline is to involve end users in the design of safety programs, initiatives, procedures or even in the development of organisational safety strategies. A common challenge shared by many safety professionals is identifying the most effective way to communicate safety procedures to employees. If you involve the employees in the development of these procedures, and they can see their contributions have been taken on board, this becomes much less of a problem. Inviting the workforce in can bring forth insights and approaches that may have never been realised from sitting behind a desk. I've had employees suggest that safety procedures should be in the form of flowcharts or decision trees rather than words on paper. I've also had employees suggest that the traditional approach of management writing procedures for the workforce should be flipped on its head by inviting the workforce to write guidance documents for management to follow.

In fact, I actually subscribe to the concept of measuring leadership decision-making as part of an organisation's suite of safety measures. Instead of organisations focussing all their safety measures at the individual worker level, I believe there is merit in seeking to measure how management decisions impact workplace safety for better or for worse. I believe a management rear-view mirror can be applied to many of the blame-inferring safety measures such as incidents, injury frequency rates, behavioural observations, safety inspections, and near misses, meaning that these measures can be applied to leaders and management in addition to (or instead of) being applied to workers. I wonder what would happen if some brave organisations asked their frontline workers to conduct behavioural observations on management.

What if the workforce were invited to initiate management inspections, measure near misses that were the result of management directions given, or report safety hazards born from poor management decisions that materialised on the shop floor? I wonder how management would feel about being involved in a safety monitoring program such as this. It may help them realise how the decisions they make trickle down and impact those at the coal face. It may also allow them to recognise how the frontline feel when management conduct safety inspections on them, pretending to be subject matter experts in frontline tasks they actually know little about instead of humbling themselves to allow for operational learning.

This is where so many organisations get it wrong when it comes to safety inspections, safety walks or safety observations. The managers or supervisors conducting these activities should recognise that they are the beneficiaries of these inspections, walks, observations and work insights – not the workers. In learning organisations, it is management who benefits from using thoughtful, inquisitive and humble questioning to build their own learning. The workers are not the beneficiaries of these activities. The workers are not learning anything new about the work they conduct by sharing their insights and feedback with management. Management conversely has everything to learn from these insights and worker feedback. Any insights gained from the frontline are a gift to management. Management is not giving anything to the workers aside from demonstrating that they actually care about the wins, losses and challenges faced by those people who are the real engine room of the organisation. Managers and supervisors (and safety teams) need to recognise that the workers are an untapped resource for learning about the environmental, organisational and job factors, and human and individual characteristics that contribute to how the safety of work plays out, and which are determining factors in the success or failure of safety programs (Health and Safety Executive, 1999). This mindset rings true for all types of work programs - not just safety programs, so the benefits of involving end users reaches far beyond the realms of workplace safety.

Your customers are the workforce and it is your job to develop a product they believe in and see the need for. You must involve the workforce in the design of your safety programs, initiatives, systems of work, and in the design of your organisations' safety strategy. You need to appeal to their intrinsic motivation rather than relying on the externally focussed motivational approaches the safety industry still clings to. Internal motivation is so much more powerful and self-actualising than external motivation, and involving your end users will build this internal motivation, encourage greater ownership of safety, and bring forth better safety outcomes by embracing and celebrating worker autonomy. As I've said a few times through this book, we need to stop doing safety *to* our people and start doing safety *with* them.

Take the development of an organisational safety strategy for example. A good safety strategy will be unmistakably hooked into the overall organisational strategy and will define a handful of safety objectives the organisation is seeking to achieve. These safety objectives should not be developed between management and the safety team behind closed doors. These safety objectives will succeed or fail on the backs of the workforce, so it makes sense to involve them in determining what they are. Safety objectives developed in an exclusionary fashion will not result in ownership by the workforce. People will not take ownership of something they perceive to have little or no control over (Lloyd, 2017). In fact, the assumptions of Reactance Theory suggest that people who perceive that choice has been taken away from them are more likely to deviate from instructions rather than abide to them (Brehm & Brehm, 2013). This often results in people doing the opposite of what they are told. If strategic safety objectives, or any safety programs for that matter, are developed without workforce involvement, and workers are simply told what to do, then organisations are more likely to witness the emergence of psychological reactance as management and safety teams hopelessly belt their heads against the wall in fruitless attempts to move the workforce towards exclusionary safety goals.

Let's take a more local level safety initiative such as a positive safety intervention program. Every organisation strives to develop a positive intervention culture where employees feel empowered and psychologically safe enough to intervene in situations where safety appears to be compromised (or at least they should strive towards this!). In designing a safety intervention program, the safety team should directly involve the end users of the program in its design. Using Learning Teams or inquiry workshops with end users will provide the most important insights into the factors of such programs that are most likely to determine if they are successful or if they will fail. You know your people, you have formed your safety coalition and you have made visible steps towards simplifying safety, so you have laid the groundwork required to engage your end users in any safety program or initiative on your organisations' agenda.

The key to success depends on the level of engagement with your customers – your end users. And whilst the power dynamics in this relationship between management and employees can feel foreign and uncomfortable for some safety professionals and managers, it is in this space where leadership grows. Acknowledging that the workforce ultimately has the power to make a safety program succeed or fail, and utilising this to build a program that meets their needs and validates their standing as key stakeholders within the organisation, is a game changer for the safety professional!

Example 3: How Brad Green Does It

Brad Green is a health and safety adjunct professor, corporate HSSE professional, thought provoker and humaniser doing great things in the US. Having worked for more than 18 years on numerous, diverse safety assignments across multiple industries, Brad prides himself on integrating humanistic approaches to foster organisational growth through learning. I asked Brad to share his insights around involving end users in the design of safety programs. Here is Brad's approach:

As a young professional, I was always keen on improving the work processes in my organisation. However, I realised that the traditional approach to creating work instructions was ineffective in capturing the intricacies and realities of the work processes. This led me to explore alternative methods to develop more effective work instructions.

One of the most significant challenges I noticed was the degree of clarity an individual tasked with creating a work instruction possessed in comprehending how work is accomplished. Their proximity to the task heavily influenced this. As an individual's distance from the work increased, their understanding of the work decreased significantly, making it increasingly challenging to prescribe how the work should be accomplished.

Unfortunately, some organisations still create work instructions without fully comprehending how the work is performed, resulting in a mismatch between the documented instructions and the practical execution of the work. This mismatch can lead to confusion, errors, and inefficiencies, causing a negative impact on the organisation's productivity and performance.

To overcome these challenges, I recognised the importance of leveraging the perspectives of the individuals responsible for executing the work. If prescribed expectations are enforced without input from the workers, they may feel powerless and may not assume ownership of the process. This lack of perceived control can negatively impact the efficacy of the prescribed work instructions. Therefore, involving the workers in prescribing how work should be done is essential, allowing them to exercise a sense of autonomy and agency over how work is accomplished.

To efficiently accumulate the valuable knowledge of our workforce, I suggest implementing a diverse team that can aid in developing well-defined processes. There are several tactics to achieve this objective, but from personal experience, I have found that facilitating a process improvement requires the facilitator to be vulnerable. The facilitator must acknowledge that they do not have absolute knowledge of the task and cannot fully prescribe the process as they do not directly engage in the activity. The facilitator must recognise the expertise of each worker involved, and to achieve an optimal outcome, each voice must be heard.

As a facilitator of various process improvements, I have incorporated the approach mentioned above, enabling me to appreciate the significance of soliciting the people's input accountable for executing the tasks. This approach has consistently demonstrated that creating work instructions based on perceived expectations can lead to conflicts, putting the individuals who face the most significant risk of injury in a precarious situation.

In summary, my approach highlights the importance of involving workers in prescribing how work should be done, allowing them to exercise a sense of autonomy and agency over how work is accomplished. By forming a diverse team and incorporating collective input, organisations can define processes that better capture the knowledge of the workforce, leading to improved outcomes.

As an HSS manager at a large chemical manufacturing company, I led implementing a uniform permit-to-work process across the organisation. Despite our site being ISO 18001 certified and receiving OSHA VPP star status, I knew this task was significant. Initially, I was concerned about changing the process, as we already had a robust permit-to-work process thoroughly embedded in our work. However, I saw this as an opportunity to gain deeper insights into our permit-to-work process from those who execute the permit-to-work process and to address any obstacles along the way.

We developed a change management plan to ensure a smooth transition and presented it to the site. As expected, we faced some resistance, and several concerns were raised. However, we invited volunteers to participate in this improvement, and in the following weeks, we received several volunteers. Through round table discussions with those using the permit-to-work process, we created various process flows for permits. We also conducted in-the-field walk-downs using the new process and identified a list of processes and procedures this change would affect.

During these interactions, I discovered that our employees genuinely understood the importance of adhering to the permit-to-work process, and this change would bring about significant efficiencies. We named permit-to-work process coaches to facilitate training and provide expert assistance after implementation. We also developed an interactive training program. The team decided that the best approach would be to limit the number of employees to eight per training to ensure everyone fully understood their roles and responsibilities.

The training room contained several easels, each with a large 3-foot by 4-foot permit-to-work station, and we went over the process flows, efficiencies, changes, improvements, and learnings discovered in the prior weeks. The interactive part of the training proved to be instrumental, and the coaches made a profound impact.

The first month of implementation did not go as intended, but that did not deter our team. We learned a lot and implemented several improvements.

In conclusion, this process and change demonstrated the importance of engaging with the people expected to execute the work, reducing the likelihood of misalignments.

The uniform permit-to-work process brought about significant efficiencies and improvements in our organisation, and our employees have embraced the change wholeheartedly.

Chapter Summary

- The workforce should be viewed as the customers of your safety programs. It is your job to develop a product they believe in and see the need for.
- Leaders and management whilst doing the best they can and with good intent, will occasionally make decisions that fail to recognise how the frontline might be impacted, ultimately increasing frontline risk.
- Managers or supervisors conducting safety walks, inspections, and work insights should recognise that they are the beneficiaries of these activities – not the workers.
- Workers are a largely untapped resource for learning about the environmental, organisational and job factors, and human and individual characteristics that contribute to how the safety of work plays out.
- Involving end users in the design of safety programs builds ownership. People will only take ownership of something if they perceive they have some level of control over it.
- The key to success depends on the level of engagement with your customers – your end users.
- Look for opportunities to co-design safety programs with those who will be expected to use them.

MICRO-PROJECT #5

Co-designing a Safety Program with End Users

If you work in the safety profession then you will be well aware of how busy things can get. Most of the safety professionals I speak with spend a considerable amount of time balancing those 'must do' pieces of work with those 'nice to have' pieces of work. For this micro-project, I want you to reach into one of those two buckets to identify a safety initiative or program that must be delivered or that you would really like to deliver to progress the safety of work within your organisation. It might be building a hazard management process, a training module, designing a safety intervention program, introducing work insights or learning teams into your organisation, or developing an engaging field safety guide. Perhaps something already exists in your organisation and you just have the sense that it is not quite landing with people the way that it should. So go ahead and pick just one safety initiative or program and let's get into it. For the purposes of this exercise, it might help to start with something small rather than a big, gnarly problem your organisation has been grappling with for some time.

STEP 1

The first step of this micro-project is to take your safety initiative and lay it out in front of you. Ask yourself who within your organisation would be affected, for better or worse, if this initiative took flight tomorrow. Would it impact people on the front line? Would this initiative require them to engage with it in some way, shape, or form? If so, which teams and which people within those teams? Would it impact supervisors, team leaders, contract managers, or project managers? What about middle management and senior management? Would there be an escalation process that pulls them in, or would this initiative require periodic reporting that they should be across? Are there any external parties who would be impacted by this change (for example, contractors or sub-contractors)? The goal of this first step is to identify your end users and anyone else whom this change might touch.

STEP 2

Now that you have done your stakeholder mapping, spend a bit of time staring into it to identify the people behind the names. With all the work you've been doing getting to know your people, you should have a pretty good understanding of at least some of the end users who exist in these teams, divisions, or groups. With a bit of luck, some of them might be members of your safety coalition whom you know are key influencers within their teams. Try to identify a representative from each team or work group whom this safety initiative could impact, and then pull that list of stakeholders together. Don't forget to include any managers, supervisors, or team leaders too!

STEP 3

Engage with these stakeholders and end users. Bring them together, float the concept of the safety initiative with them, and ask for their honest thoughts on it. You could use a learning team approach to do this, or you could just use good old-fashioned conversations. Ask questions of these end users that will help you understand if this safety initiative would result in more work for them, or what barriers they can see that might derail its success. Ask them what they think the value this initiative might add to the safety of their work. Ask them how it might land with the other people in their teams and whether there could be impacts to other areas of the business. Perhaps the optics are wrong in which case you can use the perspectives of these end users to better align the initiative with the needs of those whom it might affect. Perhaps you might be made aware of other end users you hadn't yet considered. Perhaps the initiative isn't required at all? Perhaps there is an existing process or program that can be tweaked to capture the outcomes you are seeking under this initiative. The last thing you want to do is add more work if it isn't needed. Take notes through this process. These will be really valuable to refer to as you navigate this safety initiative through the various perspectives, opinions, and experiences of all those people you have engaged with. Anything and everything you gain from this part of the exercise is a gift.

STEP 4

Having gathered and documented the perspectives of your end users, it's time to bring a handful of them together to help co-design how this initiative is going to work. You don't want too many *chefs in the kitchen* for this part – I usually ask 2 or 3 people who are well placed to represent the majority of end users to be involved. The process from here is no different to how you go about developing any workplace initiative or program with key stakeholders. Bounce things off them, ask for their expert opinions about the practicalities of the initiative, ask them to review and endorse drafts, and listen and learn from them. After all, these folks are your customers. Build the initiative around their needs.

References

Brehm, S.S., and Brehm, J.W. (2013). *Psychological Reactance: A Theory of Freedom and Control.* Academic Press.

Great Britain. Health and Safety Executive & Great Britain. Health and Safety Executive. Human Factors in Industrial Safety. (1999). *Reducing Error and Influencing Behaviour/Health & Safety Executive.* HSE Books Sudbury

Lloyd, C. (2017). *Workplace Accidents, Psychological Safety and the Crucial Role of Locus of Control.* Published on LinkedIn

11

Step 4 – Crowd Source for Safety Solutions

> **It is amazing how much you can accomplish when it doesn't matter who gets the credit.**
>
> **Harry S. Truman**

When I ask safety managers to share with me one thing that would make their workload more realistic or achievable, the answer is almost always "more resources". A bigger safety team. More bums on seats and more boots on the ground.

Safety teams are always lean. It's a common frustration among safety professionals. Too much work and not enough workers. This is a problem that I cannot see changing anytime soon. Safety teams have finite resources and little ability to influence the 'powers that be' to expand the team and recruit more solution-finders. Because despite what any well-meaning organisation prints on their billboards, shopfloor signs, or on their website, safety is not the number one priority for any organisation. Organisations do not exist to "be safe." They exist to be profitable and productive and although safety might be one of those really important values, this doesn't seem to translate to large safety teams and healthy safety budgets. Like I tell my young kids, "you get what you get and you don't get upset!"

So, how can safety teams tackle this challenge? If you are a safety professional reading this book, you probably have an annual safety plan that lists a handful of focus areas for the year, a safety strategy that presents a couple of safety initiatives or new programs to be delivered, a list as long as your arm of BAU activities and safety reporting requirements, and then somehow, you are expected to have the capacity to respond to all the unanticipated tasks that pop up on a weekly or daily basis and that are synonymous with the reactive nature of safety. How can safety teams balance all this work with so few resources? How can safety departments maximise their output to effectively influence across the entire organisation when it seems there are hardly enough people to even keep the lights on? I think the answer (or at least part of it) lies in crowd sourcing for safety solutions, which is the fourth step in the process of humanising safety.

We have established that the people closest to the work know it best and that respecting their skills and knowledge, clearing the path of safety clutter for them, co-designing safety programs with them, and humbling yourself as a safety professional or leader through operational learning is a pretty solid road to success. The last piece of the puzzle requires building a culture

DOI: 10.1201/9781032679525-14

of shared safety ownership by harnessing the skills, knowledge and perspectives of frontline workers together with the influence and decision-making ability of middle management and the commitment of the executive team to crowd source for safety solutions. This step is about developing new ways of working that decentralises power for making safety decisions from a central safety team or core management group, and gives greater autonomy and accountability to the workers. Those closest to the work who are intimately aware of the challenges that emerge on the job site, who navigate the variable nature of the work they perform, and who are the most impacted by decisions made upstream. You have taken steps to get to know your people, you have formed your safety coalition, and you have learnt how to utilise end users to help reduce that gap between 'Work-as-Imagined' and 'Work-as-Done'. Now it is time to normalise a culture of shared safety ownership by practicing a crowd sourcing approach to solving as many of those safety challenges that are inevitably going to pop up. It's time to get proactive in addressing all that reactive safety work.

This is where that safety coalition you have worked hard to form will really pay dividends. The influence these folks have throughout their teams or perhaps beyond their teams, can make your job as a safety professional so much easier, and engaging these folks to help identify solutions to safety challenges will ensure these solves make sense and will actually land with your frontline workers. How many times have you come up with a solution to a safety challenge that you and your safety team thought was brilliant, only to face resistance by the workforce when trying to implement it? This can be really frustrating – especially when you truly believe you have the answer but those stubborn frontline workers just won't engage with it. The trick to overcoming this resistance from the workforce is to engage them early in the piece and encourage them to brainstorm ideas from their perspectives and those of their co-workers. Perceived solutions serve different needs and solutions identified by management or the safety team are likely to serve different necessities than those identified by the workforce (or different levels of the workforce). Your job as a safety professional is to synthesise the knowledge, perspectives and needs of the workers – as experts in their roles – and weave it all together. This takes some pretty sharp interpersonal skills. Listening skills, communication skills, collaboration skills, conflict resolution skills, emotional intelligence and empathy. Your job is to pull this crowd of people together and harness their collective problem-solving superpowers to identify solutions that make their work easier and safer, in turn making your job as a safety professional that much easier. Let's face it - you're probably not going to get more people to join your skinny little safety team, but you surely can share the load with those influential people who represent the wider workforce, and doing this will result in solutions to safety challenges that actually make sense to the guys and girls on the frontline doing the work.

I worked at an organisation once where this crowd sourcing approach to problem solving worked really well. The organisation had the usual OH&S

Committee comprising of numerous health and safety representatives from different areas of the business, but what was different about how safety challenges were shared and solved was what the organisation labelled the 'Solutions Council.' A group of carefully elected people brought together to workshop challenges and help identify solutions (mostly related to safety), with decision-making authority formally delegated to them by the executive group. Although the focus of this council was primarily safety (every now and then challenges outside of safety were raised and worked through), you will notice that the term 'safety' was not used in the title, and this was very deliberate. Language is important when it comes to workplace safety (think about the terms safety officer, audits, investigations, inspections and all those other fault-finding words that infer blame and result in a Mexican wave of eye rolling). Within this organisation, we recognised that the term 'safety' in the naming convention for a committee or council could play into the stereotypes many people held around safety and this could subsequently stifle the types of conversations and constructive challenge we were hoping the Solutions Council would generate. So, the Solutions Council was formed and we got on with the job of crowd sourcing for safety solutions.

The council pulled together ten influential members of our safety coalition, two managers and one member of the executive team, and was chaired by one of the frontline workers from the safety coalition. We wanted to challenge the power dynamics most folks would default to by electing a chair who was not in a management role and who had the ear of the workforce. It was also a great demonstration of the value the organisation placed on the lived experiences of those on the frontline. The safety team and numerous people leaders across the business were trained in how to use work insights, and set out through our workplaces to ask workers questions such as:

- "What slows you down?"
- "What frustrates you?"
- "What is really difficult about your work?"
- "What activities could seriously injure or kill someone?"
- "If you had $100,000 to spend on safety improvements, how would you spend it?"

The safety team collated all the feedback from these work insights, categorised it, and drew out any presenting themes before sharing the summarised information to the Solutions Council. The council would select two or three challenges, problems, or themes to work through each time they came together (which was every two months). Sometimes solutions were identified during the meetings and decisions were made that allowed the required changes to progress, and other times, for challenges that were more complicated or complex, two or three council members would volunteer to take the matter on, going back to the workforce to get more information, seek wider perspectives, and invite ideas that might help solve the problem from the perspectives of the workers.

Having an executive in the Solutions Council helped keep things on track when conversations danced around perceived barriers such as time pressures or budget constraints (the council had a dedicated budget line item) and having two managers in the group helped us bridge the gap and mistrust that had inadvertently formed over time between the workforce and management. This gap was a particular challenge for this organisation, as frontline workers wanted their workplaces to be safe and the executive worked hard to provide the necersary commitment and support to raise the importance placed on safety. However, we found that it was in middle management were things sometimes came apart. Our divisional managers and team leaders were absorbing all sorts of competing objectives and priorities from downstream and upstream, and sometimes, in the rapid pace of work, safety would be de-prioritised because naturally, some trade-offs needed to be made. So, bringing a couple of managers into the Solutions Council and rotating them through every six months was a great way to build their confidence in making decisions that put our people and their safety at the centre of the discussion. We were truly transforming safety around our customers and it sent our safety climate through the roof!

I have worked in other organisations that adopted slightly different approaches to crowd sourcing for safety solutions, but as far as formal structure went, this was the most effective program I had been part of. Other variations though, included introducing safety brainstorming conversations as a standing agenda item to team meetings / existing safety committees, encouraging the safety team to consult with the workforce before presenting business cases or initiatives for approval, inviting frontline workers to executive safety meetings to share some of the safety challenges they face, feeding outputs from work insights into team improvement plans and introducing quantitative and qualitative safety metrics around work insight improvement opportunities actioned. How ever you choose to crowd source for safety, it will only build workforce trust in your safety team and benefit your overall safety program. Start small and see where it goes.

Example 4: How Zoe Nation Does It

Zoe Nation is a human factors expert, Director of On the Edge IQ, a well-recognised leader in the practical application of contemporary safety theories and an all-round great human. Zoe has over 20 years' experience designing and implementing processes that support workplace efficiency and safety across hazardous industries and is a super-engaging keynote speaker and change facilitator. She is an influential figure in the Australian contemporary safety movement and was kind enough to share with me her approach to crowd sourcing for safety solutions. Thanks Zoe!

Over time, unfortunately, we as a safety profession, have tried to have more of the answers rather than becoming increasingly humble and curious and aware that we know far less than we think. I believe a key role is having the right facilitator, someone who can provide the research and science and an objective position to help advise and align our people to make better safety decisions. However, over time I see safety professionals claim they are the expert of the topic being discussed (I'm not sure there is such a thing when it comes to the vast and ever-changing topics involved in safety). This may have been borne out of an insecurity in the profession. A need to prove our worth perhaps as we continue to have to justify numbers and have organisational restructures. Most organisations I speak to still do not align on what the role of a safety profession is and is not. The profession itself cannot agree and stakeholders have different views. As our operations have become more complex, for example, roles span over multiple different swings, worksites become more remote, technology replaces the need for a permanent role in the field, our 'customer' has become more difficult to reach, and we have found it more difficult to collaborate on safety solutions. The requests for input to safety projects typically have a tight time span on them and this means that engaging often is the step in the change process that gets left out because it takes too long.

One of the core principles of human factors in design is end-user involvement. I have been involved in designing buildings, plants, trains, equipment as well as software, policies, plans work packs and procedures for 20 years. I've been involved in cases where we have gone ahead and designed the product with the knowledge that we have based on the time and justified it because we had subject matter experts (i.e., someone who was an operator ten years ago, or a skilled designer or engineer) rather than taking time to engage with all stakeholders. And the difference between these projects, and those that have slowed down to involve the end-user is significant. An immediate consequence here is the amount of rework and the money and resources spent after we get the equipment or plant or procedure or policy. We have multiple examples and case studies to show the costs involved after the fact, however, it continues to be hard to justify and plan for at the start of a project.

Not only do we get a more fit for purpose end solution by involving those closest to the work upfront, but the act of the engagement itself during the project has many benefits. You will naturally identify and create change leaders and safety champions because instead of telling the end users what they have to make work, we are asking them to be part of the solution from the start. This means they buy into the procedure, they care about the equipment, they have pride in the solution, and this creates a better working culture, more empowerment, and more compliance with procedures. It creates a streamlined fit-for-purpose solution with an end user who believes that leaders and managers and the company as a whole, cares about their input and listens to their needs. Levels of trust increase from the outset.

The risks of not engaging those closest to the work at best might be disengagement or disempowerment or a negativity amongst the workforce, but at worst can introduce significant risk. I've seen cases where work becomes more dangerous because of a misunderstanding of the work and a design solution that is not fit for purpose. It actually introduces more complexity and creates a dangerous situation where a piece of equipment or control cannot be used, where a procedure cannot be followed, and we end up blaming worker competency or worker behaviour. But in actual fact, the design of the policy, the procedure, the system, the interface, the equipment is not fit-for-purpose from the start. They are set up to fail and those who designed the system are nowhere to be seen as they have moved on to their next project.

When I worked in the oil and gas industry in 2018, we were tasked with rolling out a new company permit system across our Australian operations within a specified timeframe and without introducing additional risks during roll out. A challenge associated with this was that we still had to continue our everyday work during the transition at the same rate as usual. Each site in Australia permitted work very differently and we had one solution design for downstream operations that we had to adapt to upstream. The organisation had been interested in contemporary safety and evolving their own safety vision and objectives for a few years, so were becoming more knowledgeable in the ideas of seeing those closest to the work as the solution. HOP and Human Factors had a part to play and some key leaders had embraced these principles and were open to managing change differently.

We engaged the workforce to leverage their perspectives through a mixture of user centred workshops, safety critical task analysis and Learning Teams. The concept of 'Work-as-Done' versus 'Work-as-Imagined' was used along with leadership direction and permission to share the current reality including rule breaches and shortcuts (promising not to punish examples of any rule breaches shared). This created a psychologically safe environment to share the current state of our permitting system.

The results of these workshops and learning teams were shocking to those in attendance, which included the permit 'SMEs' (safety professionals) and supervisors. Multiple examples of the permit system not being used in the way they had expected surprised them, and a realisation that the new permitting system was in fact going to introduce several problems and risks began to emerge. This took us back to the drawing board. During these sessions, several vocal and passionate workers took the lead and it was obvious they wanted to be involved and were our natural 'change agents'.

As expected, there was some resistance and general scepticism from the workforce, mostly due to past experiences where they had voiced concerns and shared their experiences but nothing was done in response. This was what they had become accustomed to so we had to work hard to gain their trust and make sure we provided clear expectations on when and how we would follow up with feedback. They told us how best to interact with them and we committed to that throughout the project.

There was also some initial resistance from management. A general sense of fear that we would be opening a can of worms, that we would never get alignment, that we would give them too much power, and that they would just complain. These things happened to a certain extent but not as badly as expected, because once they got a chance to vent, the natural leaders in the group started to influence their peers to align them and they could see the work we were trying to do was well intended and helpful. Managers gained valuable insights into the struggles faced by the workers. Real examples and stories became an invaluable source of learning, and the most notable benefit was that managers saw their workers wanting to solve the problems in addition to just identifying the issues. The ownership was incredibly high.

What really surprised me about this process of involving end users in the design of this permitting program and crowd sourcing to identify the solutions to the various hurdles we encountered along the way, was the willingness for the workers to give their time and effort over and above their usual jobs. They voluntarily came in during time off, were offering to be called, would stay back late when on site to try to help make the project a success. It was an overwhelming display of genuine care and passion for what started out as a corporate project. Also, there were a handful of supervisors who changed their views of their own people. Some had no idea that those involved had such skills or motivation, so it was a chance for them to see their workers in a new light.

Sure, it was not all rainbows and unicorns, as we did still have people disagree and not align with the new permit design. We made mistakes and there was confusion during deployment, but the way the team worked together on what became a common goal was refreshing. The end result was a successful change and great uptake by end users, without any significant events or close calls. Leaders were humbled and workers were empowered.

It was hard, it took longer than we had initially envisioned, and it needed a strong facilitator with knowledge of human factors who was supported by a handful of progressive leaders. We often slipped back into our old ways of *'right, that is enough engagement, let's just deliver this,'* but we had to keep forging forward, reminding everyone of the 'why' and sticking to the plan and objectives that were originally agreed. It was also an exercise in educating the teams in human performance and HOP by stealth. For example, to understand why we were simplifying some of the levels of authority and design, we had to educate everyone on the hierarchy of controls, false assurance, legal requirements and how controls are verified. The outcome was a team that had a more in-depth understanding of risk management and of each other.

Chapter Summary

- Your safety team is lean. It makes sense to use a crowd sourcing approach to solving safety challenges.
- Crowd sourcing builds a culture of shared safety ownership by harnessing the skills, knowledge and perspectives of frontline workers together with the influence and decision-making ability of middle management and the commitment of the executive team.
- Solutions identified by management or the safety team are likely to serve different needs than those identified by the workforce.
- Try to develop new ways of thinking and working that decentralises power for making safety decisions from a central safety team and gives greater autonomy and accountability to the workers.
- Use your safety coalition to learn which frontline workers you need to engage with.
- Your job as a safety professional is to synthesise the knowledge, perspectives and needs of the workers – as experts in their roles – and weave it all together.
- Expect improvement, but don't expect rainbows and unicorns.

MICRO-PROJECT #6: IDENTIFY A SAFETY CHALLENGE AND ADOPT A CROWD SOURCING APPROACH TO SOLVING IT

This micro-project is a nice one to finish on as it should demonstrate the power of co-designing safety solutions when workers feel empowered, trusted, involved, heard and valued. As with all these micro-projects, you don't need to follow this exactly as I have laid it out. Consider the context of safety in your organisation together with the relationships you hold with your end users, and just adopt the principles of crowd sourcing if that is what works for you.

STEP 1:

Identify a situation in one of your workplaces that requires attention and is presenting a problem. Ideally, for the purposes of this exercise, the situation will relate to workplace safety, however, it is not a deal-breaker if it doesn't. Crowd sourcing for solutions works across the board. The situation could be an incident or near miss, an unfavourable safety theme you have identified through crunching your safety metrics, a significant hazard or risk that has come to light, a safety process that just doesn't seem to be landing with the workforce, a safety procedure that is downright ridiculous, or a psychosocial hazard in a certain area of your organisation such as high job demands or poor organisational change management. Perhaps the safety team has already had a few 'bites at the apple' but the unresolved situation remains. That's ok – let's just get some wider perspectives on the issue and see where that takes you.

STEP 2:

Similar to micro-project 5, identify the people whom this situation affects the most. You might identify a couple of frontline staff, a line manager, someone from corporate who interacts with the data, a health and safety representative and a manager who is responsible for the operational area where this safety issue is lurking. Humble yourself by bringing them together and telling them that you are aware of this issue and that it is not something that you can fix on your own. Acknowledge the value you place on their individual perspectives and clearly state that the goal is to find a solution that makes sense to everyone whom the issue touches. Everyone in this small group of colleagues has skin in the game and their input should be valued equally.

STEP 3:

Unpack the issue. A good way to do this might be through a Learning Team format (prepare, learn, soak, improve, action) where you want to carefully frame the session by putting everyone in a learning mindset and asking about the normal everyday work surrounding the issue. Resist jumping into "fix it mode" as this discovery work is all about learning about the issue and understanding why it has become what it has become.

It is during these discovery conversations that you might begin to explore why the requirement exists in the first place. As a facilitator, you might ask questions that help uncover any dynamic trade-offs or goal conflicts giving rise to the issue, how different people or teams relate to or understand what is required of them in the context of the problem you are trying to solve, or you might identify other stakeholders you need to bring into the fold to identify a solution that meets the wider needs of the organisation and its people. This step is all about trying to understand 'Work-as-Done' rather than 'Work-as-Imagined.' Try not to rush this process – it is in these conversations that the solutions will begin to emerge, even as you resist stepping into solution mode.

STEP 4:

I find it useful to schedule a break between the discovery conversations (or session) and the solution conversations. In Learning Teams, this is referred to as 'soak time'. It might be a week, a day or two, or even just a few hours if you're up against it. Whatever the case, it is important to establish a line between discovery mode and solution mode. When you do bring everyone back together, rehash the notes you took down that equally captured everyone's perspectives and inputs as they related to discovering all the different factors feeding this issue. Then reframe people's mindsets into solution mode. This will probably feel more natural to you (and to everyone else). Ask for ideas on how to address the pain points raised during the discovery conversations and jot down some actions that will help reduce that gap between 'Work-as-Imagined' and 'Work-as-Done.' Solutions that capture the engagement of the entire group are going to be more likely to succeed, resulting in high levels of psychological safety, more effective controls, increased ownership, respect for the variable nature of work and worker autonomy, and won't result in situations where the safety team is hopelessly telling workers what to do.

12

Bringing It All together in a Human-Centred Safety Strategy

Safety. What do you think of when you hear this word? What kind of emotions does it stir?

Hopefully, after having read this book, the complexities, contradictions, dilemmas, insights, possibilities, and intricacies surrounding workplace safety excite you and generate a healthy blend of constructive and critical thinking and the emotions this loaded term, 'safety', brings forth are positive and motivating. Safety is about the humans that make up an organisation and humans are complex, fascinating, messy, beautiful and needy. But my hope is that you share my view that humanising safety doesn't seem all that difficult after all. If we adopt an eclectic approach to workplace safety theories and models, broaden our understanding of the psychology of safety, get to know our people, simplify things where possible, involve our end users when designing safety programs, and adopt a crowd sourcing approach to problem solving, then we are well and truly on our way. And in a world where workplace safety has been coerced and manipulated into a bureaucratic, administrative, dementing, arse-covering exercise by organisations in pursuit of little more than 'paper-safe' status, the only hope we, the collective safety profession have, is to restore the humanistic origins of safety together. We owe this to our people, to ourselves and to each other.

So, let me try to land this plane by bringing together and synthesising everything covered in this book into some kind of tangible, workable human-centric safety strategy that you can take and grow within your organisation. This isn't about *freezing the ocean* with a radical new approach for which your executives, leadership teams and workforce have not been primed. Rather, it is simply about making one small but significant step in the direction of humanistic safety reform. So, let's start this process by revisiting the six micro-projects of which you may have made a start on, or you might have flagged as something worthwhile to explore once you close this book. Here they are:

- Micro-Project #1: Defining Safety in Your Organisation
- Micro-Project #2: Identify and Build Your Safety Coalition
- Micro-Project #3: Get to Know Your People
- Micro-Project #4: Identify and Address One Piece of Safety Clutter

DOI: 10.1201/9781032679525-15

- Micro-Project #5: Involving End Users in the Design of a Safety Program
- Micro-Project #6: Identify a Safety Challenge and Adopt a Crowd Sourcing Approach to Solving It

Once you have worked your way through these micro-projects and flexed that organisational learning muscle, you will be ready to start making some changes to your safety strategy. I have provided a one-page template below which is just an example of how this might look and you should be able to identify where the insights, learnings and outcomes of your micro-projects feed into this template. I have provided a few prompts in italic text just to help you out.

I strongly recommend consulting end users and your safety coalition in determining your principles, desired safety outcomes and any safety initiatives/programs of work (in addition to management). So many organisations make the mistake of subscribing to the belief that safety strategies are the work of management and the safety team alone, overlooking the critical step of consulting those whom it will affect the most – frontline workers.

When it comes to brainstorming measures against your desired outcomes, a crowd sourcing approach will ensure you don't end up with low value, quantitative KPIs that mean little to the workforce and may even be perceived as 'risk-washing'. Afterall, it is likely that the performance of your safety measures will be largely dependent on how well the workforce engage with them and participate in any associated programs. Therefore, it would be foolish not involving them in determining what good measures look like.

Talk with your workers about 'Work-as-Imagined' (WAI) versus 'Work-as-Done (WAD),' about what they need from management to help them be successful, about being preoccupied with failure (i.e. by continuously asking what could go wrong), about weak signals of potential failure they experience in their work, and about what information they need access to, in real time, to help them make safer decisions. All these things should be factored into a suite of strategic safety measures. Don't fall into the trap of building measures around numbers alone. Challenge your end users and safety coalition to consider the quality of potential measures and the effect of these measures, such as whether they close the gap between WAI and WAD, whether they help build trust and care, or whether they actually help bring your organisation closer to its desired safety outcomes.

Good measures should be built around achieving the outcomes you desire and the outcomes you desire should lead to the achievement of your definition of good safety management. The way to get there and to make it stick is to humanise safety by harnessing the collective wisdom, skills, and perspectives of all the clever people around you, using four key steps to keep you on the right track. So good luck, always be kind and stay outwardly curious (Figure 12.1).

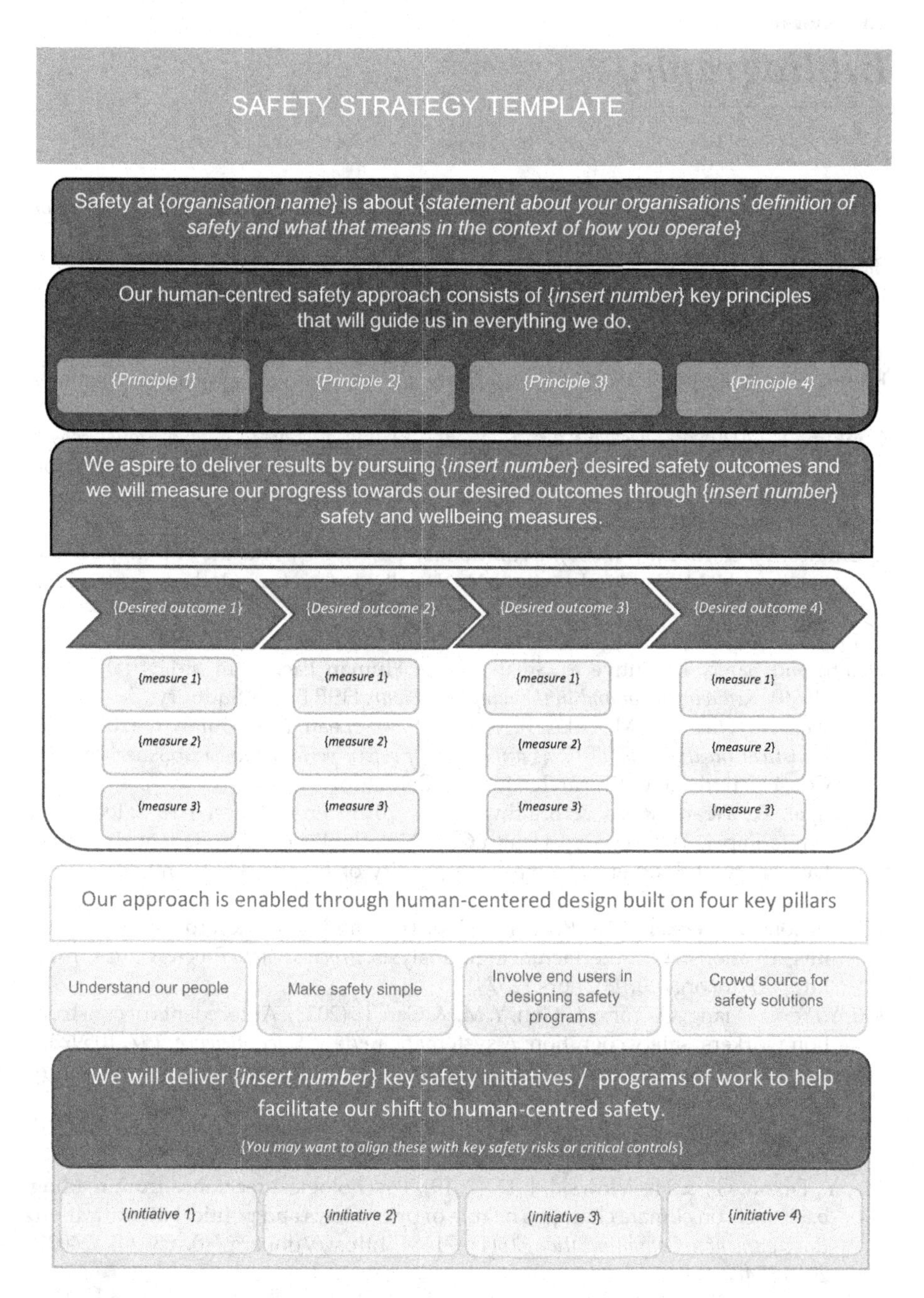

FIGURE 12.1
Human-centred Safety Strategy template.

Bibliography

Araiba, S. (2019). Current diversification of Behaviorism. *Perspectives on Behavior Science, 43*(1), 157–175. https://doi.org/10.1007/s40614-019-00207-0

Baumeister, R. F. (2020). *Social psychology and human nature.* Cengage Learning.

Bornstein, R. (2024). The psychodynamic perspective. In R. Biswas-Diener & E. Diener (Eds), *Noba textbook series: Psychology.* DEF publishers. Retrieved from https://noba.to/zdemy2cv

Brehm, S. S., & Brehm, J. W. (2013). *Psychological reactance: A theory of freedom and control.* Academic Press.

Conklin, T. (2020). *Pre-accident investigations. An introduction to organisational safety.* The Podcast. Safety Moment - Changing Behavior by Changing Behavior. https://preaccidentpodcast.podbean.com/e/safety-moment-you-cant-change-behavior-by-changing-behavior/

Cosmides, L., & Tooby, J. (2000). Evolutionary psychology and the emotions. In M. Lewis and J. M. Haviland-Jones (Eds.), *Handbook of emotions* (2nd ed., pp. 91–115). Guilford Press.

Elliott, A. (2002). *Psychoanalytic theory: An introduction.* Duke University Press.

Health and Safety Executive & Great Britain. Human Factors in Industrial Safety. (1999). *Reducing error and influencing behaviour.* HSE Books Sudbury

Hallowell, M., Quashne, M., Salas, R., Jones, M., MacLean, B., & Quinn, E. (2020). *The statistical invalidity of TRIR as a measure of safety performance.* Published by The Construction Safety Research Alliance, Colorado.

Hollnagel, E., Wears, R. L., & Braithwaite, J. (2015). From Safety-I to Safety-II: A White Paper. The Resilient Health Care Net. Published Simultaneously by the University of Southern Denmark, University of Florida, USA, and Macquarie University, Australia.

Hutchinson, B., Dekker, S., & Rae, A. (2024). How audits fail according to accident investigations: A counterfactual logic analysis. *Process Safety Progress, 2024,* 1–14. https://doi.org/10.1002/prs.12579

Liu, Y., Ye, G., Xiang, Q., Yang, J., Goh, Y. M., & Gan, L. (2023). Antecedents of construction workers' safety cognition: A systematic review. *Safety Science, 157,* 105923.

Lloyd, C. (2017). *Workplace accidents, psychological safety and the crucial role of locus of control.* Published on LinkedIn

Lloyd, C. (2020). *Next generation safety leadership: From compliance to care* (1st ed.). CRC Press. https://doi.org/10.1201/9781003051978

Ma, Y., Dixon, G., & Hmielowski, J. D. (2019). Psychological reactance from reading basic facts on climate change: The role of prior views and political identification. *Environmental Communication, 13*(1), 71–86. https://doi.org/10.1080/17524032.2018.1548369

Mcleod, S. (2023). *Theoretical perspectives of psychology (Psychological approaches).* Published by Simply Psychology.

Norsk Industri (2023). *Safety leadership and learning. A practical guide to HOP.* Published by the Federation of Norwegian Industries

Practical Psychology. (2020, June). *The Human Condition (Definition + Explanation).* Retrieved from https://practicalpie.com/the-human-condition/.

Rae, A. J., Provan, D. J., Weber, D. E., & Dekker, S. W. A. (2018). Safety clutter: The accumulation and persistence of 'safety' work that does not contribute to operational safety. *Policy and Practice in Health and Safety, 16*(2), 194–211. https://doi.org/10.1080/14773996.2018.1491147

The Journal of Humanistic Psychology. Gale General OneFile. link.gale.com/apps/pub/0DXJ/ITOF?u=ntu&sid=bookmark-ITOF. Accessed 7 December 2023.

Tooby, J., & Cosmides, L. (2008). The evolutionary psychology of the emotions and their relationship to internal regulatory variables. In M. Lewis, J. M. Haviland-Jones, & L. Feldman Barrett (Eds.), *Handbook of emotions* (3rd ed., pp. 114–137). The Guilford Press.

Turner, N., & Gray, G. (2009). Socially constructing safety. *Human Relations, 62*, 1259–1266. https://doi.org/10.1177/0018726709339863

Index

Printed in the United States
by Baker & Taylor Publisher Services

Printed in the United States
by Baker & Taylor Publisher Services